THE IMMANENT DISASTER

THE IMMANENT DISASTER
Creativity and Innovation

DR. SAMPOORAN SINGH

PUBLICATIONS
KALPAZ PUBLICATIONS
DELHI-110052

The Immanent Disaster

Rs. 750

ISBN: 81-7835-710-0

Published in 2008 in India by
Kalpaz Publications
C-30, Satyawati Nagar,
Delhi-110052
Phone : 9212729499
E-mail: kalpaz@hotmail.com

Laser Type Setting by: Rudra Computer Graphics, Delhi
Printed at : G. Print Process, Delhi

Contents

List of Figures

Acknowledgements

I am highly indebted to Dr. R. N. Sarwade, Director, Snow and Avalanche Study Establishment, DRDO, Chandigarh, for guidance of "Section I: *The Immanent Disaster: A Vision on Climate Change*". I am also indebted to Dr. Satish Kumar, Director, Terminal Ballistics Research Laboratory, DRDO, Chandigarh, for "Section II: *Creativity and Innovation*".

I used to run to these distinguished scientists as and when I was in difficulty and they gladly helped me. In fact, this treatise is a joint effort.

There are many other scientists who helped me — Pramod K. Satyawali and Saran; Ashok Kumar Abrol and Rajesh Sharma. They gave their unflinching and willing support at every step.

The whole-mankind is unwillingly and without being aware caught in a *crisis of Consciousness,* which leads to conditionings of the human psyche. I had many helpful and friendly discussions with Madame Vimala Thakar regarding dynamics of conditioned and unconditioned psyche. She is enjoying her 86th year of bliss at Mount Abu, Rajasthan.

I am grateful to Dr. (Mrs.) Kanwaljit Chopra and Shri Paramjit Singh for their collaboration in this difficult self-imposed assignment. I am thankful to S. Amarjit Singh, Editor,

Atom Science, Dagshai; Shri V. Vamanan, Editor, *Dilip,* for publishing many of my articles in their respective journals.

I am grateful to Mr. M. Natarajan, Scientific Adviser to Raksha Mantri and Director General Research and Development, Ministry of Defence, New Delhi.

Sampooran Singh

Preamble

1. Historical

The ***"World Climate Change"*** Conferences were held in 1979-1988. This was converted to the UN: IPCC [Intergovernmental Panel on Climate Change]. This has 3,000-odd scientists. It has just approved a summary of their report running to 3,000 pages of scientific reports.[1]

Hans Verolme, Director of Environmental Groups WWF's Global Climate Change Programme, said, "We have caused climate change and it is clear that we hold the solution to stop global warming in our hands".[2]

The UN: IPCC with representatives of 192-nations conference was held between December 3 and 14 in 2007 in Bali. It tackled greenhouse gas emission from deforestation to fighting global warming.[3] The conference agreed to launch a process to negotiate a new treaty that will take effect when the UN Kyoto Protocol's commitments expire in 2012.

We have to mitigate climate change, which was already causing rising temperatures, melting glaciers and extreme weather events across the world. The bone of contention was never an agreement on mandatory emission targets. The slashing of greenhouse gas emission will lead to a low-carbon society and some delegates suggested 2009 as a deadline for new climate pact.

2. The Cause of Climate Change

The UN: IPCC Chairman R.K. Pachauri in the Tribune of Nov. 18, 2007 and Nov. 27, 2007 had affirmed the warming of our climate system and linked it directly to human activity[2]. He stressed that only a few years are left to avert the worst. Some eminent scientists have shown that we have caused climate change, and it is also clear that we hold the solution to stop global warming in our hands".

The likely climate change effects are: economic from changes in rainfall patterns, debilitating loses in agriculture, the continuing retreat of Himalayan glaciers feeding vital river system and more extreme weather events. It is evident that the objective science which fosters a mere enhancement of inanimate objects of convenience cannot resolve the climatic change.

We undertook an exploration of climate change and Nature revealed Her Truth as "the fundamental direct cause of climate change" – which is the non-dual and non-conceptual frame of reference of Mind Continuum. We need dynamic interplay of ***mystical intuition*** and of ***scientific analysis***. We need to be aware of "objective and subjective observing and understanding" which bestows "peace and serenity". We have to understand and live the Wholeness of Life.

Science reveals the magnitude, geographic extent and precise locations of ocean warming. Humans are reshaping the seas through overfishing, air pollution and commercial shipping [The Tribune, February 22, 2008]. There is stronger evidence of global warming – area of glaciers reduced from 3,391 to 2,721 sq. km. between 1962-2004. They found an overall reduction of 21 percent in the overall glacial area [The Hindu, February 28, 2008]. We will have to educate the Mind which would direct humanity into the realm of supramental consciousness where the Mind is unfettered by social and economic attributes.

The fundamental crisis [direct] expresses itself as time or ego" in the external matrices. The time expresses itself as derived cause or deductive logic or "scientific analysis". The conditioning expresses as time or ego" which leads to higher conditionings of the human psyche; or in other words, there is acceleration of conditionings. The higher the conditioning the lowering the efficiency. This is bound to cut own our survival period.

The ending of "man-made" disasters is possible when the Mind attains a non-dual and non-conceptual frame of reference of Mind Continuum, or in other words, Mind having sustainable "Scientific Research and Temper".

It is pure awareness, or Consciousness, or Intelligence that heals; healing implies ending of all egos, or time, or conditionings. We have to move from knowledge to understanding, experience to perception, intellect to intelligence. We want a fundamental revolution in which the whole human race has to participate. This will guarantee survival of mankind and the planent.

Prime Minister Dr. Manmohan Singh inaugurated **"The National Plan of Action for Climate Change"** on February 7, 2008. The Plan has been linked to wide gamut of developing activities to help poverty reduction, linked Climate Change and human survival.

3. Physics and Space-Time

Physicists have come to see that all their theories of natural phenomena including the "laws" they describe, are creations of the human mind; properties of our conceptual map of reality, rather than of the reality itself .... These are creations of the mind; .... space and time are now reduced to the subjective role of the elements of a language.[4]

Relativity theory has shown that space is not three-dimensional and time is not a separate entity, both are intimately and inseparably connected and form a four-dimensional continuum which is called "space-time". Hermann Minkowski, in a famous lecture in 1908, said, "Henceforth space by itself, and time by itself, are doomed to fade away into mere shadows, and only a kind of union of the two will preserve an independent reality".[5]

All thoughts, past, inheritance, non-fact are sterile to resolve any psychological imbalances, any human challenge, because thought is a construct of symbolic-dualistic frame of reference of mind spectrum. We must jettison intellect and reason, and imbibe intuition. The terrorism, hijacking, corruption, drug trafficking and so on are constructs of thought. All droughts, floods, catastrophes, calamities, earthquakes, tsunamis, global warming are constructs of the conditioned mind-brain-body system, an inheritance of man over uncountable aeons.

4. Violence

During the last decade [1990-2000], more than 100 armed conflicts have erupted in over 70 different locations around the world. Since the end of World War II, 27 million people have lost their lives. We have built a society which is violent, and we, as human beings, are violent; the environment, the culture in which we live, is the product of our endeavour, of our struggle, of our appalling brutalities. Violence is an indication of psychological immaturity. ***Violence is a construct by symbolic-dualistic frame of reference of mind spectrum, and nonviolence is the percept in the non-dual and non-conceptual frame of reference of Mind Continuum.***

The historical conditioning of the human psyche was discussed.[4,5] Many eminent men – Lord Wavell, Radhakrishnan, Bertrand Russell, R.D. Laing, Roger Lanphear, Pupul Jayakar, VR Krishna Iyer, had also discussed the emergence of violence.

The Ending of Violence: If you sit to relax – and look at the movement of thought in utter freedom; then the energy of relaxation moves to attentiveness; and the conditioned psyche makes a quantum jump to non-dual and non-conceptual frame of reference of Mind Continuum. This ends violence.

All violence, fear, greed, acquisitions, and psychological imbalance are constructs of the conditioned psyche. We opine that to stop global warming is not in our hands as the cause of global warming is due to conditioned psyche, or inheritance. You cannot impose nonviolence on a consciousness which is violent because that violence has deep-rooted causes in the psyche. Useless these causes are transcended, there is no ending of violence.

The IPCC talked of Greenhouse Gases [CHG] such as carbon dioxide and methane released by human activities are warming the world abnormally. The hypothesis is supported by eminent scientists – say, Albert Einstein and eminent saints/ sages – say, Mahatma Gandhi, Sri Aurobindo, Vimala Thakar, Lama Govinda, John Selby, Eckhart Tolle, Lord Buddha.[6]

5. The Benediction of Nature

The *intuition, invention, discovery* provide wonderful perceptual models of "unconditioned energy", or pure perceptions, or direct responses of nature to the processes of life; but these are basically the evolution at different quantum energy potentials of indwelling consciousness.[7]

Erwin Schrodinger wrote, "If you have to face the body of a deceased friend whom you sorely miss, is it not something to realize that the body was never really the seat of his personality but only symbolically 'for practical reference?"[8] On deeper thought it seems that you face the *indwelling* consciousness every moment. Further the protagonists of quantum physics – Niels Bohr, Werner Heisenberg, Max Born and others – stated that "the object is affected by our observation. You cannot obtain any knowledge about an object while leaving it strictly isolated".[8] It implies that it is the consciousness that sees another consciousness, or every event is a play of consciousness.

(a) The essence of *"scientific research and temper"* is the conditioned and dualistic mind must engage itself in "objective observing and understanding" the whole process of life. The moment you start watching *the thinker,* a higher level of consciousness becomes activised. The holistic education must deal with the wholeness of life; and this leads to regeneration of non-dual and non-conceptual frame of reference of Mind Continuum.

(b) The mind and the brain are independent; we need restructuring of science to encompass the subjective and the objective dimensions of life; science is a human activity, it must take the ethical values in its ambient; and we have to jettison human reason and intellect, its information [knowledge] and experience, and imbibe intuition.

(c) Sri Aurobindo wrote, "The evolution of mind working upon life has developed an organization and use of matter which can no longer be supported by human capacity without an inner change. The problem is fundamental and in putting its evolutionary Nature is confronting man with a critical choice which must one day be solved in the true sense if the race is to arrive or even to survive".[9]

d) An answer to this paradox was provided by Eckhart Tolle in words, "An unprecedented shift in consciousness is indeed happening on our planet .... You are here to enable the divine purpose of the universe to unfold".[10] This enhances the awareness of the mind. World Goodwill Newsletter[11] wrote, "Humanity can be said to be on the road in mastering its subjective nature". J. Krishnamurti said, "The benediction of the timeless is waiting, the ground is ready. Can we create something that is of this benediction".[12]

6. Human Survival

George Wald, Nobel Prize in Medicine, 1967, wrote, "By now the problem is not transcendence, but survival".[13] Jacques Monod, Nobel Prize Winning biologist, wrote, "What we see opening before us today in an abyss of darkness".[14] P. Krishna, an eminent physicist, wrote, "We are going towards self-annihilation and not too much time is left now".[15] Many other scientists have talked about human survival.

7. Recent Published Papers

Man is an amphibian who lives "alternatively" in two worlds – the world of symbols and the world of matter, life and consciousness. Since 1980, I have been exploring the "conditioned psyche" and the "uncondined energy" of mind-brain-body system. I have published 43 monographs, 135 papers in Indian Philosophical Journals and 24 papers in Foreign Scientific Journals on "Science of Life" or on "Symbiosis of Spirituality and Science".

(a) I with friends initiated exploration of human mind-brain-body system. We published two monographs entitled " ***Science Education, Health and Consciousness***, Delhi; Kalpaz Publications, 2005, pp 1-354; & "***Symbiosis of Science and Spirituality***" by Delhi: Kalpaz Publications, 2006, pp. 1-396.

(b) **"Physics and Consciousness"**, Search of the Soul, 2006, Vol 3. pp. 28-41.

(c) **"An Exploration of Human Survival"**, Mumbai: Dilip, Jan/March 2006, pp. 41-45.

(d) ***"Scientific Research and Temper"***, SASE, January 29, 2007

(e) ***"A Vision on Tsunamis, Earthquakes, Snow and Avalanches and Global Warming"***, A dialogue at Chandigarh on March 2, 2007

(f) ***"Ego - Emergence and Abeyance"***, Atam Parmatam, Vol. 2, January 2008, pp. 40-43

(g) ***"Fundamental Revolution**:*, *ibid.*, April 2008 [in press]

8. Concluding Comment

(a) All floods, droughts, catastrophes, calamities, earthquakes, tsunamis, snow & avalanches, global warming, are not due to nature's wrath against mankind, but are constructs of the conditioned mind-brain-body system or inheritance of man over uncountable aeons.

(b) Man has to learn to make a quantum jump from intellect to intuition [insight]. It appears that when a packet of energy [violence-Unmanifest-violence] stored in the realm outside space-time and this is governed by superluminal channel; but this release is located in space-time matrices, which appears as earthquakes. All calamities are due to transformation [fronzen time] to time.

(c) Without being aware of it and without being rigorously systematic about it, man has started living in a man-made cosmos. Albert Einstein, in a prophetic vision, stated that we need, "objective observing and understanding". This results in jettisoning the symbolic-dualistic frame of reference of mind-spectrum and make a quantum jump to non-dual and non-conceptual mode of Mind Continuum. The Pure Consciousness becomes self-referral, and this is called subjective science, this will help to deeply understanding higher quantum jump of Awareness or Consciousness, which resolves all human challenges. This is called fundamental research programme and is direct expression of the unconditioned human psyche; and is known as *mystical intuition"*. The derived crisis is direct expression of nature and called *"scientific analysis"*. Our knowledge of "natural clamities" are basically equivalent to that of a high school student.

(d) The ending of "man-made" disasters is possible when the Mind attains a non-dual and non-conceptual frame of reference of Mind Continuum; or in other words, a Mind having sustainable, "Scientific Research and Temper".

(e) We have to move from knowledge to understanding, experience to perception, intellect to intelligence. We want a fundamental revolution, a psychological revolution, or inward growth, in which the whole human race has to participate. This will guarantee the survival of mankind and the planet.

(f) What the Buddha proclaimed about the imperative need to *know* the mind, to *shape* the mind, and to *liberate* the mind has never been more true than in the Age of the Atom.

References

1. R.K. Pachauri, in *The Tribune of Nov. 18, 2007;* and Dec. 15, 2007.
2. *Ibid,* in *The Tribune of Nov. 27, 2007.*
3. Fritj of Capra, *The Tao of Physics,* London: Bantam Books, 1988 [Second Ed.], pp. 277, 147, 154.
4. Sampooran Singh, *The Survival of Mankind and Quantum Jump from Violence to Peace,* New Delhi: Books India International, 1999, pp. 223-225, Cover Jacket.
5. *The Millennium World Peace Summit of Religions and Spiritual Leaders,* New York, August 2000.
6. Sampooran Singh and Dr. Mrs. Kanwaljit Kaur, *Symbiosis of Science and Spirituality,* Delhi: Kalpaz Publications, 2006, pp. 372-374; 345-380.
7. Sampooran Singh, *Scientific Research and Temper,* SASE, Chandigarh, January 29, 2007, pp. 23-24.
8. Erwin Schrodinger, *What is Life? and Mind and Matter,* Cambridge University Press, 1967, pp. 134, 135.

9. Shri Aurobindo, in Kishore Gandhi, *The Odyssey of Science, Culture and Consciousness,* New Delhi: Abhinav Publications, 1990, p. 41.
10. Eckhart Tolle, *The Powers of Now,* Hodder, 2004, pp. xiii [preface], vii.
11. World Goodwill Newsletter, 2006, No. 4.
12. J. Krishnamurti; *A Biography,* in Pupual Jayakar, San Francisco: Haper & Row Publishers, 1986, p. 380.
13. Kishore Gandhi [Ed.], in *Evolution of Consciousness,* New Deilhi: National Publishing House, 1983, p. 11.
14. Jacques Monod, in *Synthesis of Science and Religion,* Mumbai: The Bhaktivedanta Institute, 1988; in New York: Vintage Books p. 171.
15. P. Krishna in *Education, Science and spirituality,* USA: The University of California and Santa Barbara, 1993, p. 12.

Abstracts

1. The Act of Human Living

Objective science follows objective analytical procedures and on the role of the intellect as distinct from intuition and feeling. It has no capacity to resolve the calamities and psychological challenges. The UN: IPCC must engage itself in the fundamental *[mystical intuition]* and the derived *[scientific analysis]*. It is the whole [fundamental] that determines the behaviour of the past [all calamities] and the IPCC must deal with both. We want a fundamental revolution or inward growth in which the whole human race has to participate.

2. Human Revolution

Albert Einstein, in a prophetic vision, said, "Mankind needs objective observing and understanding. Both observing and understanding are embedded in the non-dual and non-conceptual frames of reference of Mind Continuum. Observation without an observer is attentivity. The movement of reaction-free perception, unpolluted perception is the very action of observation. Meditation is a way of living where relationships are movement of relaxation and no tensions and pressures. Spirituality proceeds from the awareness of the whole, and to analyse the particular. It explores the instruments of perception.

3. Global Warming

We discuss "Human Crisis" and the "Planet Crisis" and to explore technologies to tackle "global warming". The human crisis is called fundamental crisis *[mystical intuition]* and the planet crisis is called derived crisis *[scientific analysis]*. The Clean Development Mechanism [CDM] was promoted by the governments for over two decades, but the results have been minimal. Eminent scientists – Paul Davies, Max Born, Hideki Yukawa, Fritjof Capra – have discussed technology to resolve the present paradox. Only the Intelligence of Love and Compassion can solve the problems of life.

4. A Vision on Tsunamis, Earthquakes, Snow and Avalanches and Global Warming

Without being aware of it and without being rigorously systematic about it, man has started living in a man-made cosmos. Mind has erected the objective outside world of natural philosopher out of its own stuff. The world extended in space and time is but our representation.

Every human being is throwing energy of violence in the atmosphere. When it reaches a critical level, it travels deep into the earth [as superluminal channel] and gets recorded in timeless realm. At appropriate moment, it gets released as time, as a calamity. History records only violence (time), it cannot record the timelessness.

Man has to learn to make a quantum jump from intellect to intuition [insight]. It appears that when a packet of energy [violence-Unmanifest-violence] stored in the realm outside space-time and this is governed by superluminal channel; but this release is located in space-time matrices, which appears as earthquakes. All calamities are due to transformation [frozen time] to time.

All floods, droughts, catastrophes, calamities, earthquakes, tsunamis, snow & avalanches, global warming, are not due to nature's wrath against mankind, but are constructs of the conditioned mind-brain-body system or inheritance of man over uncountable aeons.

The ending of "man-made" disasters is possible when the Mind attains a non-dual and non-conceptual frame of reference of Mind Continuum; or in other words, a Mind having sustainable, "Scientific Research and Temper".

5. Immanent Disaster

We unequivocally affirm that objective science alone cannot resolve the "human-induced climate change, and we need *mystical intuition* and *scientific analysis*. We present "scientific analysis" that the conditioning of the human psyche is increasing at a rapid rate, and this conditioning is bound to cut down our survival period.

Human-induced climate change adds new stress – as nervous tensions and chemical pressures. We want new relations between man and nature, and between man and man. We need a dynamic interplay of objective science and subjective science [spirituality].

The only solution to ending of "human-induced climate change" is higher quantum energy potential of Awareness/ Attentivity/Consciousness. It is imperative to initiate wholistic [holistic] education at all levels at the earliest.

6. The Nature of God, The Universe, The Climate Change and Man

If humanity is to survive on this earth, we will have to undertake education of the conditioned and the Unconditioned Mind; the ego or the I-consciousness of the mind spectrum; and to meet the challenges that create and perpetuate disharmony, chaos and violence. The UN : IPCC must engage itself in a fundamental challenge (***mystical intuition***) and derived challenge (***scientific analysis***). It is a crisis in our Consciousness.

Prime Minister Dr. Manmohan Singh has inaugurated "The National Plan for Action for Climate Change" on February 7, 2008.

This dialogue presents transformation, or mutation, or revolution of the whole psyche through Truth or Love. Its expression on human survival is presented. Man is miniaturized cosmos. He is a limited being. His perception is that of schoolchildren. Man is not the measure of himself. No man from outside can make you free. He has to educate himself to perceive Nature's benediction. It is Nature's "grace" that can resolve all Climate Change or Global Warming, and this will lead mankind to survival.

7. The States of Mind

Man has inherited a violent through-structure which has led to a challenge of Climate Change. Though cannot transform and bring transformation in the human psyche.

The movement of the Mind is the movement of the total human race. It mainly deals with "the nature of God, the universe and man". We have to educate the Mind for symbolic-dualistic frame of reference to non-dual and non-conceptual frame of Mind Continuum. It implies to observe and understand intuition, invention and discovery; and to move through Silence to Meditation. A non-subjective action is a communion with Nature, which bestows "peace and serenity".

8. The Ending of the Climate Change

Thought is mechanical and repetitive, so it is sterile in bringing any transformation in the psyche. Loss of gift of foresight, loss of intuition, mediocre living add to enhancing conditioning and loss of humanity.

Observing and understanding happens in a non-dual and non-conceptual frame of reference of Mind Continuum. Where there is a cause there is an ending to that cause. We need direct perception of the fact and mutation of the psyche. We have to educate the Mind so that it revels in higher potential of Awareness/Consciousness. It is the Universal Consciousness, or the Life Field, or the Intelligence that ends all challenges of Climate Change, and guarantees human survival.

9. The Ending of the Human Challenges

The mind excluded the Subject of Cognizance [Consciousness] and constructed objective world of psychological imbalances at about million years ago. These imbalances are called conditionings or inheritance.

Albert Einstein, in prophetic vision, stated that the "objective [including subjective] observing and understanding" where both are embedded in a non-dual and non-conceptual frames of reference of Mind Continuum. This resolves or ends all psychological imbalances – violence, fear, greed, acquisition and so on.

Consciousness is non-local and infinite, therefore Immortal and Eternal. The Divine Universal Consciousness is the truth of the universe. Man recognizes the truth of the immanence and transcendent potentials of each human consciousness.

Man is a miniaturized cosmos. Our knowledge about God is but the knowledge of schoolchildren. Everything is governed by immutable laws. All men are put up as "toys" by God. Everyman has to educate himself to perceive the immutable laws of nature and be ready to perceive the benediction of nature.

10. The March of Objective Science

Without being aware of it, it discards the "subject of cognizance" and is called "objectivation". It constructs violence, fear, acquisition, motives, derives, and these are called "conditionings".

The mind has the capacity to make a quantum jump from dualistic frame to non-dual and non-conceptual frame, and this is called intuition. The intuition has the capacity to creativity [wisdom]. If a non-dual frame of mind thinks of a commercial activity, it falls from non-dual frame to dualistic frame, and is called "innovation".

Mankind needs a higher level of consciousness, a quantum jump is consciousness - from knowledge to understanding, experience to perception, reason to intuition, and intellect to Intelligence.

11. Creativity and Innovation

Creativity implies discovering something, which nobody has hitherto known about, or inventing something new. It is a discovery of truth in nature hitherto unknown to man.

The moment you think that you want to use the invention [intuition) for a commercial activity, the quantum jump potential falls to dualistic and conceptual mode, he loses quantum jump and falls to innovation.

The scientist has to repeat the whole process of "intuition, invention and discovery" to regenerate himself.

12. Science Research and a Scientific Temper

Man is a miniaturized cosmos. It explores "Science Research and a Scientific Temper". It also explores innovation in science and in man. President A.P.J. Abdul Kalam addressing a seminar at the University of Delhi on December 31, 2004 discussed "creative leadership". In another seminar at IIT, Delhi, on March 31, 2005 he said that India should "become a continuous innovator and creator of science and technology -intensive products". M.M. Sharma on July 14, 2004 said, "We as a nation will have to innovate to survive". R. Chidambaram also advocated, "Generation of innovations and also support systems". We intuitively perceive that the survival and excellence of man is embedded in "self-education", which flows to discovery and which Itself guides for its innovation.

All great discoveries have arisen from the realm of the timelessness, so *scientific enquiry is not a cerebral activity and gathering knowledge [information] but an expression or a quest of the Timelessness, the Divinity, the Life Field. Innovation is to use the discovery to a commercially useful process / product. Discovery is fundamental and innovation is a derivative.*

Innovation is embedded in the depth dimension of man but the tool to be followed is scientific research.

The survival of human race is embedded in our faculty of intuition [insight]; and insight self-expresses Itself as innovation.

13. Cause of Our Failure in Innovation

J Krishnamuri stated, " India has lived all these centuries on ideas. She has to move from a life based on ideas, which are non-facts, to seeing of fact only." It is the "ego" at the materialistic level or dualistic level which is the primary causes of *our failure in the field of innovations. Your consciousness is shared by all humanity. So the fact, the cause is shared by mankind.*

14. Rationalism and the March of Reason

The intellect is invariably dualistic and it constructs "cultural belief structures" or "ego-structures". When I am aware of the limitations of the intellect, the mind itself makes a quantum jump from dualistic state to non-dual and non-conceptual state's and this state knocks down the ego-structure.

The innovation is in the time field, while the mutation is in the timeless realm; and mutation is invariably synergistic to innovation.

15. The Act of Living

Man has conditioned his psyche in the last 2.5 *million years. He has started living in the distorted present, which leads to nervous tension and chemical presence which is guiding him to self-destruction. He has lost the receptivity, the sensitivity. This is leading him to self-destruction.*

He is now aware that the task of burning the age-old tide of human ambition and aggression in very difficult; and all his efforts have failed so far. Albert Einstein, in prophetic words, suggested that the man has to learn "objective observing and understanding". We have to learn about subjective research.

16. Consciousness, Creativity and Innovation

All creativity - intuition, invention, discovery and intelligence - are expressions of consciousness.

Creativity and innovation are two pillars on which the progress of man hinges - the creativity is a pure perception of Nature, which is the internal, the infinite, the immutable, the truth; and the innovation is a construct of the conditioned mind which is the past, the finite and the transient.

Creativity is Nature's response to our challenges; and innovation refers to use of Nature's response at a lower level of awareness, which converts to a commercially useful process / product. There is a need for sustainable development of new techniques to beat global competitors.

Innovation deals with materialism, so it divides the mind to a subject and an object, which leads to dualism. The dualistic patterns lead to violence, conflict, and emergence of "*ego*" or "*I-ness*"*, and psychological imbalances. The ego damages the innovation process.*

When the innovation is observed by the intense energy of creativity [insight], the ego and psychological imbalances go to abeyance and the whole system reverts to scientific temper. The survival of the human race is embedded in our faculties of creativity and innovation.

The aim of this is to initiate an open dialogue between scientist-mystics working in the field of "*cognitive science*" or "*physics of Consciousness*"*.*

17. What is Life?

Life is infinite and eternal. It is self-generated and self-sustained. Creativity is the discovery of truth in nature hitherto unknown to man, it is the nature's response to our challenges, and it deals with intuition, invention and discovery.

When a scientist wants to use his invention for technology - which is to lead for something commercially useful - the mind moves to materialistic activity, or *dualism, or thought-ego or lower quantum energy potential of consciousness. This is called "innovation". President AP.J. Abdul Kalam, R. Chidambaram also discussed creative leadership and innovation. M.M. Sharma wrote,* "We *as a nation will have to innovate to survive".*

The scientist will have to encounter "objective observing and understanding" to regenerate his mind and imbibe, objective truth. The objective truth will lead to the "modified innovated structure", that will have the capacity to beat global competition. This exploration is bound to achieve a higher potential for the Defence R&D for "Innovation".

18. The Battle of Human Psychological Imbalances

Creativity is the discovery of truth in *nature hitherto unknown to man, it is the nature's response to our challenges, and it deals with the wholeness of life. It leads to intuition, invention and discovery.*

We intuit that the exclusion of the Subject of Cognizance in science is due to a sudden encounter of the "intense energy of insight", which results in enhancing the potential of the object. This leads to that the moral, spiritual and other dimension is normally pushed out of science. So the nature of science is violence and destruction.

The psychological imbalances, or cultural belief structure, or ego-structure is also called innovation. This article discusses the resolution of in novation to creativity. This leads to ending of violence, aggression, terrorism, corruption, and so on.

Innovation refers to an idea of "intuition or Invention" to a commercially useful process / product. It is a materialistic process, an economic activity. The failure in discovering the truth is attributed to thought-ego, dualism and psychological imbalances. It deals with objective values and is devoid of subjective dimension.

The innovator has to re-explore the "innovated structure" through "objective observing and understanding". This leads to non-dual and non-conceptual Mind Continuum, which heals the innovated structure.

Holistic education deals with a quantum jump from knowledge to understanding and from intellect to intelligence. The problem facing mankind is to become aware with a new dimension of consciousness, which will generate a new relationship between man and nature, man and man, and man and environment.

We intuit that this exploration is bound to enhance [about 3 times] the R & D potential for "innovation". This also leads science to subjective and objective dimensions. The education explores the wholeness of life.

19. The Survival of Mankind?

This article explores the survival of mankind. We have to restructure objective science [outer dimension] and subjective science [inner dimension]. It discusses the scientific research and temper. The "Principle of Objectivation" excludes the Subject of Cognizance from the domain of nature that we endeavour to understand.

All objective facts - colour and sound, hot and cold - are lacking in our world model. We intuit that the mind-brain-body system has both subjective and objective dimension. This strongly suggests that in our exploration of Consciousness, we have to consider both subjective and objective dimension.

We must jettison intellect and reason, and imbibe intuition. We must make a quantum jump from dualistic living to non-dual and non-conceptual frame of *reference* of *Mind Continuum, which is called scientific temper. The key to this quantum jump is "objective and subjective dimensions of observing and understanding". We have to be simultaneously aware of the wholeness of life. And we call this as Intelligence, or Wisdom, or Superluminal dimension of Life.*

20. The Benediction of Nature: The Energy of Intelligence

Many eminent scientists - say, Erwin Schrodinger, Charles Sherrington, B. Spinoza, C. G. Jung and so on - have stated, "Mind has erected the objective outside world of the natural philosopher out of its own stuff". This implies that all psychological imbalances and all calamities [including tsunamis, earthquakes, snow and avalanches, global warming] are constructs of the conditioned human psyche.

Knowledge belongs to the mind, not to consciousness, Pure perception belongs to consciousness. Inner awareness is essential for observing and understanding. Human personality depends on "inner conscious awareness".

Albert Einstein; in Prophetic words, said that we need "Objective observing and understanding" to achieve, "peace and serenity". This energies the "Energy of intelligence", so the benediction of the timeless is waiting, the ground is ready". The creativity {intuition, invention and discovery] is the energy of Intelligence.

We have to restructure science so that it takes in its realm both the objective and the subjective dimensions of life. The restructured science has the dimension:

[Science-Education-Ethics-Health]

The ethics imply to take in its ambient the sensual and trans-sensual dimensions of life. A living in the present, the now heals and living in the past and duality suppresses man's perceptive sensitivity and enhances disease. We must be aware of the mutation of the whole. It is the higher level of awareness that heals. It is the energy of Intelligence that heals.

All floods, catastrophes, calamities, earthquakes, are not due to nature's wrath against mankind, but are constraints of the conditioned mind-brain-body system for inheritance of man over uncountable aeons.

The energy of Intelligence arises from the Silence, so it is non-subjective and non-objective. It is a flame of attentivity, it is sensitivity. The energy of Intelligence transcends all human challenges and bestows freedom and peace.

21. Nature's Benevolence of the Hidden Potentials of Consciousness

Man is an amphibian who lives alternately in two worlds - man-made social cosmos [psychological mode of mind] and a Creative cosmos [intuition, invention, discovery], which is permeated by Intelligence or Wisdom.

If man is aware of the psycho-dynamics of man-made social cosmos, then nature, out of its benediction, does grace and bestows creativity. The creativity banishes all violence, strife, fear, nervous tension and chemical pressures, and man revels in perfect freedom and peace, truth and love.

"Objective observing and understanding" stimulates the higher potential of Consciousness that heals. We have to explore the human subjective dimension, or inner realm or inner conscious awareness. We have to restructure our 'objective' science and take in its fold both the objective and the subjective dimension of life.

Nature, in its benediction, is full of harmony and order, non-violence and non-conflict, love and compassion, perfect sustainable freedom and peace.

Section-I

The Immanent Disaster : A Vision on Climate Change

1

The Act of "Human Living"

1. Modern Physics and the New Science

Man is an amphibian who lives alternately in two worlds – man-made social cosmos [psychological mode of mind] and a Creative Cosmos [intuition, invention, discovery], which is permeated by Intelligence or Wisdom.

The civilization in which we live today is the product of the human mind, known as modern science. Science aims at nothing but making true and adequate statements about its objects. The scientist only imposes two things namely truth and sincerity, imposes upon himself and upon other scientists.

Science has certainly contributed considerably towards the progress and well-being of mankind in many fields. Ironically, however, increasing visible side effects of industry and technology confront us with such grave dangers as environmental hazards, shortage of natural resources, and nuclear war. The latest challenges introduced are global terrorism, organized crime, corruption, drug-trafficking, hijacking and so on. These indicate an accelerated erosion of human values and ethical standards leading inevitably to the nullification of the achievements of modern science.

The explosive progress of material technologies; search of technical solutions of our challenges; combination of open-mindedness and following of abstract concepts; the logical and rational aspects; military strategy as weapons of mass destruction [WMD's] have led objective science to violence. The whole question of moral, spiritual, and other dimensions is normally pushed out of science. The resolution of our problems at thought level threatens to stabilize and even to destroy human civilization. These aggravates current socio-economic-political tensions. Accompanied by raging ecological problems, these tensions tend to push the world ever closer to self-destruction.

Ilya Prigogine, a Nobel Prize in 1977 wrote, "The behaviour of a reality is unrelated to our beliefs, our ambitions, our hopes .... A triumph of human reason turned into a sad truth. It seems that science debased everything it touched."[1] Alexander Koyre wrote, "[Science has made] a world in which, though there is a place for everything, there is no place for man".[2] Voltaire wrote, "Everything is governed by immutable laws... everything is prearranged... everything is a necessary effect...".[3] H. K. Dua, in a Tribune debate, *"India at Sixty"* highlighted the shortcomings and failures in the last 60 years. He wrote:

The Constitution itself is under strain. Parliament, the executive, the judiciary, political parties and other institutions engaged in the governance of the country of over a billion people are not functioning as they ought to..... The quality of democracy has sadly declined.[4]

This speaks of the limitations of modern science, and its failures to resolve the psychological challenges.

The discoveries of modern physics tell us that thought is sterile to resolve any psychological challenge, because thought is a construct of symbolic-dualistic frame of reference of mind spectrum. The resolution of this paradox is as under:

(a) We have to restructure objective science [outer dimension] to subjective science [inner dimension]. It discusses the scientific research and temper. The "Principle of Objectivation" excludes the Subject of Cognizance from the domain of nature that we endeavour to understand.

(b) We intuit that the mind-brain-body system has both subjective and objective dimensions. This strongly suggests that in our exploration of Consciousness, we have to consider both subjective and objective dimensions.

(c) We must jettison intellect and reason, and imbibe intuition. We must make a quantum jump from dualistic living to non-dual and non-conceptual frame of reference of Mind Continuum, which is called scientific temper. The key to this quantum jump is "objective and subjective dimensions of observing and understanding". We have to be simultaneously aware of the wholeness of life. And we call this as Intelligence or Wisdom, or Superluminal dimension of Life.

(d) Albert Einstein, in a prophetic vision, stated, "Objective observing and understanding" stimulates the higher potential of Consciousness that heals. We have to explore the human subjective dimension, or inner realm or inner conscious awareness. We have to restructure "objective" science and take in its fold both the objective and the subjective dimensions of life.

(e) Maurice H. Wilkins, Nobel prize in medicine, 1962, wrote,"The whole question of moral, spiritual, and other dimensions is normally pushed out of science. So something in the nature of science often does lead people towards destruction."[5] He added, "[Science] has produced a narrow-minded way of thinking amongst scientists."

(f) Victor F. Weisskopf in the *Introduction* to Hideki Yukawa monograph entitled *"Creativity and Intuition"* wrote,

"They also demonstrate how difficult is the task of turning the age-old tide of human ambition and aggression .... [there is] the large gap between the superhuman task and the frailties of human endeavors...."[6]

(g) All great discoveries of science have arisen from the realm of the timeless, so scientific enquiry is not a cerebral activity and gathering information, but a quest of the timelessness, the Divinity, the Life Field. A selfless unfaltering devotion to truth and use of perception for man's uplift – higher hierarchical expression of consciousness – are the only vocation of scientists.

2. Education and Value

We want a revolution, psychological revolution, in which the whole human race has to participate and every man is an equal partner. This revolution will regenerate or rejuvenate so that it perceives the new-dimensions of an integral and all-inclusive consciousness, that can guarantee human survival along with all fauna and flora.

In the words of Ramana Maharishi, "Illiteracy is ignorance and education is learned ignorance. Both are ignorant of the true aim .... The true aim is the non-dual made of knowing, while being literate or illiterate concerns the dualistic mode of knowing. Ignorance, in other words, is ignorance of the non-dual and non-conceptual mode of knowing, which would reveal the universe to the Mind only."[7] This suggests that we have to restructure education.

Maurice H. Wilkins wrote, "I agree with Einstein that the sort of scientific education we have now has produced a narrow-minded way of thinking amongst scientists, so that they give no proper attention to the moral and psychological dimensions."[5] This suggests that we have not paid attention to the development of the educational structure through our institutions. Wilkins added that, "Science as it is today does not tell us very much about the real nature of human beings."[5]

Recently many eminent scientists – Werner Heisenberg, Edwin Hubble, Max Planck, G.F. Chew, Bertrand Russell, Lincoln Barnett, Max Born and so on – have concluded that we cannot impart value education at reason or intellect level".[8] Wilkins stated, "The question of good and bad, of human virtue relates to every human activity. How therefore can science, being a human activity, be separated from all virtues and values."[9] D.S. Kothari stated, "The core of rational Science is the outer world and the core of value is the inner world. Science and values are complementary".[10] Education and values are inseparable.

3. New Paradigm of Education in UK/Europe

The Scientific and Medical Network, UK, which has over two and a half thousand members, has started a new Educational Initiative, called the Blaker Foundation "It aims at a transition from our evolving but materialistically based culture to a broader spiritually inspired understanding of the world". Blaker (1999) put it succinctly, "Our science of today cannot stop the spiritual advance, but can only delay it for a while, for such delays will automatically drag in their wake calamitous consequences, including geological disasters". There is a silent renaissance going on in UK and Europe. In some universities, UK has started conferring M.A., M. Phil and Ph. D. degrees. It is exploring a new educational structure "to allow teaching within a spiritual and holistic context in the United Kingdom." This vision is a pointer to a fundamental revolution in restructuring of education. We in India are in deep slumber. The fault lies with our scientists who have lost the creativity.

4. The Great Calamities/Catastrophes/Disasters/ Earthquakes/Tsunamis/ Global Warming

Fritjof Capra, an eminent physicist, discussed 'natural' disasters in 1998 that are caused in part by human-induced global climate change. He stated that Hurricane Mitch took 10,000 lives; the catastrophic flood in the Yangtze River in

China caused more than 4,000 deaths, and Bangladesh suffered floods killing 1,400 people. Capra added that "the rising tide of natural catastrophes over the past is a clear indication that the climate instability caused by human action is increasing".[11] Janet Abramovity, World Watch Institute, wrote, "The unnatural disasters were made more frequent or more severe due to human action".[12] Capra concluded that "The principal lesson to be learned is that the cause of most of our environments and social problems are deeply embedded in our economic systems."

On December 26, 2004, huge tidal waves were generated by a strong earthquake hit coastal areas of India, Sri Lanka, Indonesia and Thailand and flattened everything that came in its way. The death toll crossed 2,83,000 with many missing, injured and homeless.

A New Vision on Earthquakes: A great calamity – known as Bihar Earthquake descended on India on January 15, 1934. Nearly 25,000 people were swallowed up in the bowels of the earth in the twinkling of an eye, and tens of thousand had been rendered homeless. Mahatma Gandhi's teachings in response to Bihar Earthquake are under:

Gandhiji wrote, "Behind even this calamity there is a divine purpose that works for the good of all humanity .... this earthquake is a divine chastisement sent by God for our sins .... that not a blade of grass moves but by the divine will .... For me, there is vital connection between the Bihar calamity and the untouchability campaign.[13]

He further wrote, "Let us derive the lesson from this calamity that this earthly existence is no more permanent than that of the moths we see every night dancing round lights for a few minutes and then being destroyed.[14] He added, "This absolutely unthinkable affliction in Bihar – your sins and my sins – are responsible .... this affliction has come upon us because of this atrocious sin of untouchability". He advocated that "We have to open the book of life. The root meaning of education is drawing out."

To me the earthquake was no caprice of God nor a result of a meeting of mere blind forces ...He and His Law are one. This Law is God. Anything attributed to Him is not a mere Attribute. He is the Attribute. His is Truth, Love, Law, and million things that human ingenuity can name .... For God is the Law. But I submit that we do not know the Law or the Laws fully, and what appear to us as catastrophes are so only because we do not know the universal laws sufficiently .... Visitations like droughts, floods, earthquakes and the like, though they seem to have only physical origins, are, for me, somehow connected with man's morals .... With me the connection between cosmic phenomena and human behaviour is a living faith that draws me nearer to my God, humbles me and makes me readier for facing Him.[15] Mahatma Ghandhi wrote "Nonviolance is the law of our species as violence is the law of the brute."

Violence is a construct by symbolic-dualistic frame of reference of mind spectrum, so it is the conditioned energy. Nonviolence is a percept in the non-dual and non-conceptual frame of reference of Mind Continuum so it is the non-conditioned energy. A non-violent mind revels in freedom and peace.

To sum up all droughts, floods, catastrophes, calamites, earthquakes, tsunamis are not due to nature's wrath against mankind, but are constructs of the conditioned mind-brain- body system, or inheritance of man over uncountable eons. The responses by saints-seers, philosophers-mystics of East and West – Albert Einstein, Mahatma Gandhi, Sri Aurobindo, Vimala Thakar, Lama Govinda, John Selby, Tolle Eckhart, Lord Buddha – seem to support the view. Man is a miniaturised cosmos so all cosmic phenomena and human behaviour happens to be intertwined and interrelated.[16] An understanding gives us the feeling that man divorced from nature could not be happy and man's power to resist nature is hopelessly puny.

Tsunamis: A tsunami is a wave or series of waves in the ocean that can be hundreds of miles long and have been known to reach heights up to 34 ft [10.5 m]. These "walls of water" travel as fast or faster than a commercial jet. The massive December 26, 2004 tsunami travelled 375 miles [600 km] in 75 minutes. The walls of water are capable inflicting massive damage along coastal lands.

Paul Davies, an eminent physicist, wrote, "There is unexpected harmony and coherence of nature ....nature is astonishingly clever in the way it gets things to work .... These are the products of exceedingly ingenious design ...the universe has a purpose.[17]

Global Warming: Billions and billions of people are throwing the energy of violence throughout the waking state and the dream state of mind in the environments. When a conditioned mind-brain-body system throws such a packet of energy in the atmosphere, it is stored up in the atmosphere and from there it goes deep down the earth and gets stored there. Physics tells us that energy cannot be destroyed, so it is locked up deep in the earth, frozen in time. If the stored up energy cannot be sustained it gets unlocked and gets dissipated as an earthquake or volcano eruption. When the energy of violence attains a critical mass, it explodes as earthquake, but this criticality is not amenable to any measurement. If there is any *"fault"* in the earth it appears as earthquake which has an elemental dynamics of its own. The earthquake is defined as an explosion when the energy of violence has a critical mass and coupled with the dynamics of the fault in the sea had.

The packet of violence energy is locked up deep in the ocean frozen in time. The energy of violence is fundamentally discontinuous, so between any two consecutive quantum-energy packets, there is the unmanifest, it is basically "violence-Unmanifast-violence" packet. Nature due to its ingenuity permits the energy from the Unmanifast to the manifest. When the packet of energy violence-Unmanifast-violence reaches a critical level, it explodes [information transfer] in the real outside space-time and this is governed by superluminal channels. It travels deep in water and gets locked up in some "fault" in the ocean bad. The fault with locked up energy explodes in the timeless as violence and gives rise to tsunami.

Scientists are now monitoring events in the melting Greenland ice cap [Paul Brown, the Hindu, September 13, 2007], and [Daniel Howden, the Tribune September 10, 2007] that the breaking of glaciers are triggering earthquakes, and this break-up appears to be catastrophic. It is only future research of IPCC ["UN: Intergovernmental Panel on Climate Change] that will indicate the future.

UN Secretary General Ban Ki-moon, (The Sunday Tribune, Nov 18, 2007] stated that he made a trip to see ice shelves breaking up". He said "But these are even more terrifying because they are real. Hans Verolme said, "We have caused climate change, and it is also clear that we hold the solution to stop global warming in our hands". We opine that to stop global warming is not in our hands as the cause of global warming is due to conditioned human psyche, or inheritance. You cannot impose nonviolence on a consciousness which is violent because that violence has deep-rooted causes in the psyche. Unless these causes are dominated there is no ending of violence.

Maurice H Wilkins wrote "The logical and rational is a very important element in science. But emphasis tend to make people not notice the essential role of intuition."[18] "Science is the research of technical solution to our challenges," It is an intellectual attempt and has no capacity to resolve the paradox.

We intuit that objective science has no capacity to resolve the challenge of global warming or other calamities.

5. Resolution of Human Paradox

Capra paraphrased an old Chinese saying, "Mystics understand the roots of the *Tao* but not its branches; scientists understand its branches hut not its roots. Science does not need mysticism and mysticism does not need science, but men and women need both. Mystical experience is necessary to understand the deepest nature of things and science is essential for modern life".[19] We call a dynamic interplay between mystical intuition and scientific analysis. We call the mystical approach to understand the cause of global climate change as "fundamental" and steps to end global warming and other challenges as "derivative." We have penned a monograph entitled "Dynamic Interplay of Science and Spirituality'[20] — the IPCC may publish it at the highest priority.

Einstein wrote, "The chief motivating force behind all technical effort ought to be a supreme concern for the destiny of the human race" It is the Cosmic Consciousness that governs the mutations of the mind-brain-body system. It activises the Intelligence which expresses as order, harmony, freedom, nonviolence and peace. It is a unique service to mankind. It has the potential to steer the ship of humanity from confrontation to sharing, from annihilation to survival. And if man escapes his responsibility to work for mutation of his psyche, it will be at his own peril and that of India and mankind.

6. Concluding Comment

Maurice H. Wilkins wrote, "Science makes progress largely by a combination of two aspects: open-mindedness and the following of abstract concepts .... Science [follows] analytical procedures and on the role of the intellect as distinct from intuition and feeling".[5] The objective science has no capacity to resolve the calamities and psychological challenges. All violence is a construct of the conditioned mind-brain-body system and calamities are expressions of human violence.

Pachauri said [The Tribune, Nov. 27, 2007] that, "The IPCC had unequivocally affirmed the warming of our climate system and linked it directly to human activity". The whole mankind under the guidance of UN: IPCC must engage itself in the fundamental [mystical intuition] and the derived [scientific analysis]. It is the whole [fundamental] that determines the behaviour of the parts [all calamities] and the IPCC must deal with both. We have to learn to understand deeply the Awareness or Consciousness, as that can only resolve the human challenges, the human calamities. In this is embedded the human survival and excellence. This is the only vocation of the I PCC and other scientists.

We want a fundamental revolution, or psychological revolution, or inward growth in which the whole human race has to participate. The "objective observing and understanding" leads to regeneration or rejuvenation of human Mind so that it perceives the new dimension of an integral and all-inclusive Consciousness or Awareness. This will accelerate the self-unfolding of Truth and the Divine Grace. This will generate freedom, nonviolence and peace, which will end all derived challenges. This will guarantee the survival of mankind and the planet.

References

1. Ilya Prigogine and Isabelle Stengers, *Order out of Chaos,* London: Bantam Books April 1984, pp. 5,6.
2. Alexander Koyre, *ibid.,* pp 35,36.
3. Voltaire, in *Dictionnaire Philosophique,* Paris: Garnier, 1954, in Ilya Prigogine, p. 257.
4. H.K Dua, *Editor in Chief, The Tribune,* August 12, 2007.
5. Maurice H. Wilkins, in *Synthesis of Science and Religion,* Mumbai: The Bhaktivedanta Institute, June 1988, pp. 30,35.

6. Victor F. Wasskopf, in *Introduction to Hideki Yukawa Creativity and Intuition,* Tokyo: Kodansha International Ltd. 1973, 13.
7. Ramana Maharshi, in Ken Wilber, *The Spectrum of Consciousness,* USA: The Theosophical Publishing House 1982, Third Quest Book, p. 110 .
8. Sampooran Singh Dr. (Mrs) Kanwaljit Kaur, Paramjit Singh, *Global Education and Human Survival,* Delhi: Sanbun Publishers, March, 2002, pp 46, 35-36.
9. Maurice H. Wilkins, *op. cit.,* p. 37.
10. D.S. Kothari, *Fifth Bhai Parmanand Memorial Lecture,* New Delhi: December 1983, pp. 3, 5.
11. Fritjof Capra, *The Hidden Connections,* New Delhi: Harper Collins Publishers, 2003, pp. 183 – 184; in Sampooran Singh and Dr. (Mrs.) Kanwaljit Kaur, Symbiosis of *Science and Spirituality,* Delhi: Kalpaz Publications 2006, pp. 363-364.
12. Janet Abramovity, in Fritjof Capra, *ibid.,* p. 184, in Sampooran Singh et al., *ibid.,* 364.
13. Harijan, 02.02.1934, *The Hindu,* 24.01.1934 and the *Hindustan Times,* 25.01.1934.
14. Harijan, 02.02.1934, *The Hindu,* 26.01.1934.
15. Harijan, 16.02.1934. *The Collected Works of Mahatma Gandhi,* Govt of India, The Publication Division. No LVII, pp. 164.165.
16. Paul Davies, *Superforce,* New York: Simon and Schuster, 1984, pp. 224, 223, 232, 243.
17. Sampooran Singh and Dr. (Mrs.) Kanwaljit Kaur, *Symbiosis of Science and Spirituality,* Delhi: Kalpaz Publications, 2006, pp. 372-373, 344.
18. Maurice H. Wilkins, *op. cit.,* p. 31.
19. Fritjof Capra, *The Tao of Physics [Second Edition],* New York: Bantam Books, 1988, p. 297.
20. Sampooran Singh, *Dynamic Interplay of Science and Spirituality,* unpublished.

2

Human Revolution

1. Revolution or Transformation or Mutation

Man is an amphibian who lives *"alternately"* in two worlds – the world of symbols and the world of matter, life and consciousness. There must be a revolution – but not a revolution based on an idea, or a concept, or a thought-experience. You cannot impose an order on top of disorder. You cannot impose non-violence on a Consciousness which is violent because that violence has deep-rooted causes in the psyche. Unless these causes are eliminated there is no ending of violence.[1]

Revolution or transformation or mutation is not in the future, can never be in future. It can only be *now,* from moment to moment. J. Krishnamurti wrote, "Seeing the false as the false and the true as the true is transformation, because when you see something very clearly as the truth, that truth liberates. When you see that something is false, that false thing drops away".[2]

Mutation is a social change in the quality of the mutant, in the substance of the mutant. It is not a peripheral change. It is a radical change in the whole of the mutant. Such a mutation is possible in the brain through insight, not through knowledge, not through accumulation. That insight is the ability to see a truth directly, a direct perception of the truth and education must foster that.[1] So we must not only emphasize knowledge, but emphasize **insight** – the possibility of insights.

This article explores the truth, the love, which is an alternative way of living love is not different from truth. Love is that state in which the thought process, as time has completely ceased. Without love, revolution has no meaning. It also discusses the different phases of revolution or transformation.

2. Civilization is a Process of Conditioning

Man had an evolutionary origin through uncountable aeons. The mind slowly developed to reason or intellect at about 2.5 million years ago. The origin of thought is embedded in deep antiquity and slowly it developed to reason. Fritjof Capra wrote that "Evolution of human nature was virtually completed some fifty thousand years ago. Since then the human body and brain have remained essentially the same in structure and size".[3]

Wilder Penfield, a renowned Canadian brain surgeon, wrote "Our Minds and brains are not at all the same. It is the mind which experiences and it is the brain which records the experience...."[4]. To suppose that consciousness or the mind has location is a failure to understand neurophysiology".[5] It Mind is consciousness and consciousness is an energy. All psychological imbalances – violence, fear, greed, anger, and so on – are stored in the brain; and this is called "ego", "I-consciousness". Ego has no factual reality at all. It has a conceptual reality. It is an ideational reality. The superstructure of concepts [ego] has been superimposed on thought and created by the human race.

3. "Objective Observing and Understanding"

If one sees the limitations of the conditioned psyche, the mind makes a quantum jump to a higher state of attentivity. In other words, the symbolic-dualistic mind is transcended and it takes the role of non-dual and non-conceptual Mind Continuum. The mind is regenerated to higher Attentivity/ Awareness.

Albert Einstein, in a prophetic vision, said, "Mankind needs objective observing and Understanding".[6] It implies to see non-subjectively, it is to observe without any evaluatory comparisons or judgment creeping in. It implies to see without naming, identifying, without emotionally reacting. Observation is perception without psychological involvement, without any kind of reaction. Observation without the observer is the state of wholeness of perception and wholeness of responses. It leads to an enquiry where the receptivity, the sensitivity is exposed to a consciousness. There is neither subject nor the object. It is a subject-free and object-free attentivity.

4. Ending of "Ego", "I-consciousness"

Life consists of movement and non-movement, motion and non-motion, action and non-action. These two together make the wholeness of life. There is shear motionlessness. No thought moves, no feeling, no sentiment, no emotion is moving. Motionlessness implies ending of all nervous tension and chemical disturbance; identifications with concepts, and so results in ending of ego.

We use "Consciousness for observing". We see that the movement of thought-structure is not a voluntary movement at all; it has nothing to with his will, nothing to do with his desire, it has its own independent movements. It is independent of my volition, it is independent of the ego. Cerebral movement [thought-experience-memory] becomes irrelevant to learning.

I observe the movement of thought – not to judge, neither to accept nor to reject, not to act upon them, not to change them. I look at the movement of thoughts in utter freedom. The thinker and the thought are not separate. There is nothing like an individual mind, there is nothing like an individual consciousness. There is racial consciousness and a racial momentum of knowledge-experience, and there is a sense of great freedom within.[7]

5. The Dimensions of Silence

In silence, the non-personal energy, non-individualized energy of Awareness or Intelligence functions. In silence you get reduced to nothingness – nobodyness. We have to observe subjectively, to observe without any evaluatory comparisons or judgments creeping in.

Observation without an observer is attentivity. The movement of reaction-free perception, unpolluted perception is the very action of observation.

The occurrence of seeing or hearing or breathing is a movement in the sphere of non-duality. There is no subject-object division and no movement born of the tension of duality. The movement of organic intelligence in the complexity of cosmic life is a movement born of non-duality and conducted in non-duality.

Awareness is choiceless. It is unrelated to the human past including knowledge and experience – it is called choiceless Awareness.

6. Meditation

It is allowing the total conditioned mind to go to abeyance, and relax completely. It is a total relaxation of the conditioned mind. It is a way of living where relationships are movement of relaxation and not tensions and pressures. A movement of thought is equal to tension in your whole nervous system; any movement of emotions and sentiments is equal to chemical turbulence in the body. Silence is a tension-free and pressure-free state of your being. Being in silence is being in the wholeness of your being.

Meditation is a way of observing in total tranquility the stream of thought whereby the thought quietens down, enabling one to experience the silence. Feelings of separation and duality disappear and one becomes conscious of the hidden dimension [Intelligence].

7. Spirituality

It is a science that treats the whole of life – the macrocosm, the microcosm, the world of infinite small [say, atoms and sub-atoms] and the individual – as its field of investigation, experimentation and exploration. It proceeds form the awareness of the whole, and to analyse the particular.

The spirituality explores the instrument of perception. In spiritual science the emphasis is upon the act of perception, the quality of instruments employed for perception, the quality of consciousness behind the perception.

Spirituality means the quietening down, the cessation of our thinking, our conditioned stream of thought. It is a way of life that teaches us to get into contact with the inner silence in yourself.

Both science and spirituality follow the law of non-authoritarian approach – an approach not conditioned by the authority of the past.

8. Love

Thought emerged from the symbolic-dualistic frame of reference of mind spectrum; it gets recorded in the brain. It is invariably coupled with ego, so it constructs concepts. When violence, greed, fear, anger, possession and other psychological imbalances are transcended, there is love. Love alone can transform the present madness and insanity in the world – not system, not theories. Love or Truth is that state in which the thought process, as time, has completely ceased. Psychology has not given the direction of living for Truth, or Freedom, or Love.

9. Life

Thought-experience-memory developed to reason and language Intellect, the mind and its whole contrivance is the creation of the center as the "me", the "ego". The ego has a conceptual but not an objective reality.

The intellect slowly developed to intuition, which is non-dual and non-conceptual frame of reference of Mind Continuum. The non-dual frame slowly developed to Intelligence, where the Intelligence is the Awareness of the wholeness. This results in understanding and a spontaneous response of the whole.

Life is a mysterious wholeness. It is homogeneous, undifferentiated and dimensionless. The ecological crisis that exists in the world today is the result of some contamination in the basic perception of life.

10. Concluding Comment

Human beings are born for enjoying unconditional inner freedom. It is only free human beings who can cooperate with one another.

Every human being has to learn to live in truth, towards the homogeneous wholeness of life, towards the mystery of life. We have the faith that the mysterious inter-relatedness shall help when help is needed. That faith is necessary for conducting an enquiry on the subtle realm of consciousness, for conducting an enquiry into the **invisible** and the **infinite.**

We have to offer our life for that psychic mutation [transformation, revolution] to take place, we might be creating a living cell of a new life style, a new way of living. We might be setting into the orbit of human consciousness, a spark of love and compassion. We embark upon that enquiry on behalf of the whole human race.

It is the higher hierarchical level of Awareness/ Consciousness that can resolve all human problems, and bestow on man freedom, peace and love. Every existence in governed by immutable law of Nature. This is the benediction of nature, a free gift bestowed to every man.

References

1. P. Krishna, *Education, Science and Spirituality,* California: Institute for Higher Education, Research and Dialogue 1993, pp. 15, 14.
2. J. Krishnamurti, *The First and Last Freedom,* London: Victor Gollancz Ltd., Thirteenth Impression, January 1977, p. 286.
3. Fritjof Capra, *The Turning Point,* London: Wildwood House, 1982, p. 325.
4. Wilder Penfiled, in Deepak Chopra, How to Know God, London; Rider, 2000, pp. 215-217.
5. *Ibid;*, in Fred Alan Wolf, San Francisco: Harper and Row, 1981, p. 244.
6. Kenneth Brecher, *Albert Einstein: 14 March 1879 – 18 April 1995, Nature,* vol. 278, 15 March 1979, p. 215.
7. Vimala Thakar, ***Silence***, Holland: Bookfund Vimala Thakar, Balaricum; and *Through Silence to Meditation,* Ahmedabad: Vimal Prakashan Trust. 1989.

3

Global Warming

1. The Immanent Calamity

The UN: IPCC [Intergovernmental Panel on Climate Change] has 3,000-odd scientists.[1] The representations of 192-nations conference met between December 3 and 14, 2007 in Bali and approved a summary of their report running to 3,000 pages of scientific reports. The IPCC confirmed that "only a few years are left to avert the worst".[2] The warming of our climate system and linked it directly to human activity. Hans Verolme said, "We have caused climate change and it is clear that we hold solution to stop global warming in our hands". Ban Ki-moon, UN Secretary General, visited ice-glaciers and stated that, "I come to you humbled after seeing some of the most precious treasures of our planet – measures that are being threatened by humanity's own hand ... But They are even more terrifying, because they are real".

The explosive progress of material technologies; search of technical solutions of our challenges; combination of open-mindedness and following of abstract concepts; the logical and rational aspects; military strategy as weapons of mass destruction [WMD's] have led objective science to violence. The whole question of moral, spiritual, and other dimensions is normally pushed out of science.

We discuss "Human Crisis" and the "Planet Crisis" and also explore technologies to tackle climate change crisis, or resolve the challenges of global warming. The human crisis is called fundamental crisis *[mystical intuition]* and the planet crisis is called derived crisis *[scientific analysis]*. We suggest that the IPCC may examine the fundamental crisis and derived crisis as both are essential for human survival. In other words, we need the basis of both scientific evidence and moral need.[3]

2. The Human Crisis and the Planet Crisis

The clean development mechanism [CDM] was promoted by the government far over two decades for reducing energy use and air pollution, but the results have been minimal. We need to limit concentrations of CO_2 to 400 ppm [parts per million] or below [compared with 280 ppm in the pre-industrialised era and 380 ppm today]. Obviously the CDM alone cannot resolve the challenges.

Many eminent scientists-philosophers-educationists have divided the human challenge and the planet challenge into three parts – *Basic, Fundamental* and *Derived* crises, and are discussed as under:

Basic Crisis: Homo sapiens emerged at about 2.5 million years ago. The thought's origin is embedded in deep antiquity. Physics has confirmed that all the concepts we use to describe nature are limited, that they are not parts of reality, but creations of the mind; parts of the map, not of the territory. All concepts arise from symbolic-dualistic mind; and these have no originality. These are recorded in the brain and called "conditionings". The conditionings construct fear, violence, anger, acquisition, inheritance. Man has failed so far to resolve the conditionings.

The conditioned-mind is mechanical, repetitive and has no originality, no newness. It is a form of insanity. It is disharmonious and disordered. It lives in the past memory. Psychologically we go through agonies, tortures, uncertainties, despair, and a great sorrow.

Hideki Yukawa, a Japanese Nobel laureate, wrote, "Man has things in him that are unknown even to himself .... Every individual has in him that cannot be simply defined".[5] The objective science or rational thought cannot resolve our conditionings. All thoughts, past, inheritance, non-fact are sterile to resolve any psychological imbalances.

Fundamental Crisis: Albert Einstein, in a prophetic vision, stated, "Objective observing and understanding" stimulates the higher potential of Consciousness that heals.[6] It implies that "observation" transcends the symbolic-dualistic frame of reference and wakes up the non-dual and non-conceptual frame of reference of Mind Continuum. The non-dual frame of reference is invariably synergistic to symbolic-dualistic frame of reference of mind spectrum.

Observing [seeing], listening [hearing], touching, taste is direct; thinking, thought, inheritance is indirect.

One has to learn to observe. When attention that is evaluatory perception is sustained, the subjective momentum of identifying, comparing goes to abeyance. Observation implies not to judge, neither to accept, nor to reject, not to act upon them, not to change them. I look at the movement of thought in utter freedom. When it is seen that there is social momentum of knowledge, experience and thought, and there is nothing like an individual ego, the self, the me – there is a sense of great freedom.[7]

Life consists of movement and non-movement, motion and non-motion, action and non-action. These together make the wholeness of life. There is sheer motionlessness. No thought moves, no feeling, no sentiment, no emotion is moving. When the thought, the inheritance, the past, the non-fact transcend, then nothing is left behind; and this is called the *"silence"*. Silence is a tension and pressure-free state of your being.

Meditation is a way of living where relationships are movements of relaxation and not tension and pressure. This stimulates the energy of Intelligence, which has the capacity to resolve all human challenges. Meditation is a direct expression of Consciousness or Awareness.

Derived Crisis: A great calamity – known as Bihar Earthquake descended on India on January 15, 1934. Nearly 25,000 people were swolled up in the bowels of the earth in the twinkling of an eye, and tens of thousand had been rendered homeless. Gandhiji wrote, "Behind this calamity there is a divine purpose that works for the good of humanity".[8] All such calamities which are caused by "nature" are called derived crisis.

Violence is a construct by symbolic-dualistic frame of reference of mind spectrum *[basic crisis]* and non-violence is the percept in the non-dual and non-conceptual frame of reference of Mind Continuum *[fundamental crisis]*. When the fundamental crisis is expressed by nature, it is called "derived crisis". It is direct expression of nature.

The IPCC talked of Greenhouse Gases [GHG] such as carbon dioxide and methane released by human activities are warming the world abnormally. All derived crisis are constructs of the human conditioned mind and supplemented by deforestation and pollution. The hypothesis is supported by eminent scientists – say, Albert Einstein and eminent saints and sages – say, Mahatma Gandhi, Sri Aurobindo, Vimala Thakar, Lama Govinda, John Selby, Eckhart Tolle, Lord Buddha.[9]

3. Nature's Benediction

Paul Davies, an eminent scientist, wrote, "The laws which enable the universe to come into being spontaneously seem themselves to be the product of exceedingly ingenious design. If physics is the product of design, the universe must have a purpose, and the evidence of modern physics suggests to me that the purpose includes us".[10] He added, "Nature is astonishingly clever in the way it gets the things to work". Man is a miniaturized cosmos, so "man divorced from nature could not be happy and man's power to resist nature is hopelessly puny".

Nature, in its benediction, is full of harmony and order, non-violence and non-conflict, love and compassion, perfect sustainable freedom and peace, and free of all psychological imbalances and contradictions. We have to learn to live according to the immutable laws of nature.[11]

We have to explore the fundamental crisis *[mystical intuition]* and the derived crisis *[scientific analysis]* as both are governed by the immutable, ingenious design of nature.

Max Born, an eminent physicist, wrote, "Clever, rational ways of thinking are not enough. The danger of mass slaughter ... can only be overcome by moral conviction, by a determination to replace national prides and prejudices with human love".[12] The Pugwash Conference [1962] wrote, "We reassert our conviction that the goal of full disarmament and permanent peace is realistic and urgent .... We call upon scientists everywhere in the world to join us in this task".[13]

Hideki Yukawa stated, "Science is a manifestation of man's effort to open the unknown world that lies before him, to discover new possibilities".[14]

4. The Great Calamities/Catastrophes/Disasters/ Earthquakes/ Tsunamis/ Global Warming

Fritjof Capra quoted some great calamities.[15] He added that , "The rising tide of natural catastrophes over the past decade is a clear indication that the climate stability caused by human action is increasing". Janet Abramovetz added, "The natural disasters were made more frequent or more severe due to human action".[16] Capra concluded that the principle lesson to be learned is that the cause of most of our environmental and social problems are deeply embedded in our economic system. We conclude that all natural calamities are embedded in the human system.

Billions and billions of people are throwing the energy of conditioned violence [symbolic-dualistic frame of reference] throughout the waking state and the dream state and this gets stored in the environment. This storing of energy of violence is further supplemented by the human activity – deforestation and pollution. The IPCC has "unequivocally affirmed the warming of our climate system and linked it to human activity". When this stored violent energy meets a *fault* [at earth's surface or deep in the ocean] it gets stored there. Paul Davies stated that nature is astonishingly clever" and it releases the stored violent energy as earthquake or tsunami. All activities are direct expression of nature.

5. Concluding Comment

Every human being is throwing energy of violence in the atmosphere. When it reaches a critical level, it travels deep into the earth [as superluminal channel] and gets recorded in timeless realm. At appropriate moment, it gets released as time, as a calamity. History records only violence (time), it cannot record the timelessness.

All floods, droughts, catastrophes, calamities, earthquakes, tsunamis, snow & avalanches, global warming are not due to nature's wrath against mankind, but are constructs of the conditioned mind-brain-body system or inheritance of man over uncountable aeons.

The ending of "man-made" disasters is possible when the Mind attains a non-dual and non-conceptual frame of reference of Mind Continuum; or in other words, a Mind having sustainable, *"Scientific Research and Temper"*.

Man has to learn to make a quantum jump from intellect to intuition [insight]. It appears that when a packet of energy [violence-Unmanifest-violence] stored in the realm outside space-time and this is governed by superluminal channel; but this release is located in space-time matrices, which appears as earthquakes. All calamities are due to transformation [frozen time] to time.[17]

The future is going to be low-carbon society, and we have to bring out a transition through *mystical-intuition* and *scientific analysis.*

The new relationship of the great calamities – including earthquakes, tsunamis, global warming – is that instead of rupture and opposition between man and nature, there is growing coherence of our insight – *[mystical intuition]* and *[scientific analysis];* or in other words, we want a dynamic interplay of science and spirituality. The aim of "natural calamity" [derived crisis] is to remind the human race that it must jettison the antagonistic attitude and learn to live in "harmony and order".

Every event/organism is governed by the immutable laws of Life, by Cosmic Consciousness, by Life Field. Everything is prearranged .... everything is a necessary effect. The truth of man and the truth of nature is the "Oneness of Life". Every expression of nature – both in the fundamental crisis and the derived crisis – is interwoven and interlinked with the true nature of man. Man is the miniaturized cosmos.

Only the Intelligence of Love and Compassion can solve the problems of life. That Intelligence is the only instrument that can never become dull, useless. To perceive and to live in the "Immeasurable State of Nature" [Divine or Intelligence] is the greatest vocation in the world. Love is not different from truth. Love is that state in which the thought process as time, has completely ceased. Where love is there is transformation. This is the only vocation of mankind.

If the conditioned mind voluntarily goes to non-action, the Cosmic Intelligence takes over the change [mutation] of our being, and all disasters and catastrophes go to abeyance and man lives in *Sat-Chit-Ananda* or *Love-Freedom-Peace; or Truth-Consciousness-Bliss.*[18]

References

1. R.K. Pachauri, in *Times for India,* October 13, 2007; & *The Hindu,* December 16, 2007.
2. *Ibid., The Sunday Tribune,* November 18, 2007; & *The Tribune,* November 27, 2007.
3. Sudhir Chella Rajan, *The Hindu,* December 20, 2007.
4. Fritjof Capra, *The Tao of Physics [Second Edition],* London: Bantam Books, 1988, pp. 147, 277.
5. Hideki Yukawa, *Creativity and Intuition,* Tokya: Kodansha International Ltd., 1973, pp. 206, 9.
6. Albert Einstein, 14 March, 1879 – 18 April, 1955, in Kenneth Brecher, *Nature,* Vol. 278, 15 march 1979, p. 215.
7. Vimala Thakar, in *Through Silence to Meditation,* Ahmedabad: Vimal Prakashan Trust, 1998, pp. 19, 25, 27, 13.
8. M.K. Gandhiji, *Harijan,* 02.02.1934; 16.2.1934; 20.4.1934.
9. Sampooran Singh and Dr. Mrs. Kanwaljit Kaur, *Symbiosis of Science and Spirituality,* Delhi: Kalpaz Publications 2006, pp. 372-374; 379.
10. Paul Davies, *Superforce,* New York: Simon and Schuster, 1984, pp. 243, 232.
11. Sampooran Singh, *An Exploration of Human Survival, DILIP,* Jan.-March 2008, pp. 41-45.
12. Max Born, in Heidki Yukawa, *op., cit.,* pp. 189-190.
13. Declaration by *The Tenth Pugwash Conferecne, 1962,* in Hideki Yukawa, *op., cit.,* p. 191.
14. Hideki Yukawa, *op., cit.,* p. 205.
15. Fritjof Capra, *The Hidden Connections,* New Delhi: Harper Collins Publications, India, 2003, pp. 183-84; in *op., cit.,* [reference 8], pp. 363-364.

16. Janet Abramovetz, in [reference 9], p. 364.
17. Sampooran Singh, *A Vision on Tsunamis, Earthquakes, Snow and Avalanche, and Global Warming,* A Dialogue at Chandigarh on 02.03.07.
18. *ibid., Preamble,* December 25, 2007.

4

A Vision on Tsunamis, Earthquakes, Snow and Avalanches and Global Warming

1. Introduction

The physical-biological-psychosocial evolution over billions and billions of years have shown that there is harmony and order in all events/existences. There is love and compassion and nature's benediction flows every where. But strange it seems, the disasters [often called natural disasters], or catastrophes, or calamities take a very heavy toll of human life and property year after year. The United Nations estimates that in the past two decades, nearly three million lives have been lost to natural disasters and some 800 million affected by them. Every year, at least four cyclones of varying intensity hit parts of India's nearly 6,000-km-long coastline. Besides, earthquakes of varying magnitudes rock northern India.

Albert Einstein stated, "We know nothing about it at all [nature of God, the universe and man]. Our knowledge is but the knowledge of schoolchildren .... we shall know a little more that we do now. But the real nature of things – that we shall never know, never".[1] As an illustration, astronomers reported a cosmic explosion just across the Milky Way from Earth gave off as much energy in one-tenth of a second as the sun does in 100,000 years", which was observed on December 27, 2004.[2]

Another illustration is that "the Union of Concerned Scientists [UCS]" – over 1670 scientists, including 104 Nobel laureates – issued "World Scientists Warning to Humanity" on April 1993 in words:

A new ethic is required – a new attitude towards discharging our responsibility for caring for ourselves and for the earth. We must recognize the earth's limited capacity to provide for us. We must recognize its fragility. We must no longer allow it to be ravaged. This ethic must motivate a great movement, convincing reluctant leaders and reluctant governments and reluctant peoples themselves to effect the needed changes.[3]

The latest international report confirming the "Unequivocal" nature of global warming.[4] We do not know the real nature of things about *Tsunamis, Earthquakes, Snow & Avalancches.*[5]

The article is an exploration of man-made [natural] disasters, or catastrophes, or calamities by *"logical"* extension of the *unpredictable* and *known* extension of facts, but with a limitation as we have not been able to advance any experimental evidence of the phenomenon. We quoted several eminent scientists. This is also supported by world saints/seges.

The hypothesis is supported by eminent scientists – say, Albert Einstein – and eminent saints/sages – say Mahatma Gandhi, Sri Aurobindo, Vimala Thakar, Lama Govinda, John Selly, Eckhart Tolle, Lord Buddha. Every word uttered by these saints/sages is in the timeless realm. Nature's every expression is in the timeless realm and free from any violence, conflict, contradiction; and imbibe freedom and peace. It suggests strongly to me that there is no scope for a great calamity, or catastrophe, or a Tsunami, except human failure to deal with the crisis of transcending violence and conflict.

President A.P.J. Abdul Kalam visited St. Petersburg on May 25, 2005 and called for "pooling of scientific knowledge and resources between the countries to enhance capabilities for the prediction of earthquakes, Tsunamis, and other national disasters".[6] The two sides have formulated about a dozen joint projects.

2. Tsunami

A tsunami is a wave or series of waves in the ocean that can be hundred of miles long and have been known to reach heights up to 34 ft. [10.5m]. These walls of water travel as fast or faster than a commercial jet. The massive wave on December 26, 2004 traveled 375 miles [600 km] in 75 minutes. That is 300 mph [480 kmph]. These walls of water are capable of inflicting massive damage along coastal lands.

On December 26, 2004, a massive underwater earthquake off the coast of Indonesia's Sumatra island rattled the earth in its orbit. The quake measuring 9.0 on the Richter scale in the largest since 1964. Dozens of aftershocks with magnitude of 5.0 or higher occurred in the following days. The death toll has reached 2,83,000 and many communities suffered devastating property damage.

Energy of Violence: Violence is an indication of psychological immaturity. Physics concludes that violence is a construct by symbolic-dualistic frame of reference of mind; and non-violence is a direct percept of the non-dual and non-conceptual frame of reference of Mind Continuum. The non-dual frame is synergistic or at a higher quantum energy potential to symbolic-dualistic frame. We have to understand and learn a quantum jump from violence to non-violence, from reason to intuition. And in this is embedded our survival and excellence.

Billions of men are constructing violence at every instant and throwing it in the external environments. We know nothing as to when this energy becomes critical and how it gets dissipated.

Planck propounded the fundamental theorem that "higher frequency means higher energy". Energy of peace is at the highest frequency, and energy of violence is at the lowest frequency. Energy of peace is in communion with the laws of nature. It is also called Intelligence, it is immutable and eternal. Energy of violence is mutable, it is a construct of the conditioned psyche, or man-made construct. It is antagonistic to the laws of nature.

Observing [seeing], listening [hearing], understanding and learning, action are direct; while thinking and psychological inheritance is indirect. Nature's expression is invariably direct.

The energy of violence is fundamentally discontinuous, so between any two consecutive quantum energy packets, there is the Unmanifest – it is basically "violence-Unmanifest-violence" packet. Nature due to its ingenuity permits the energy from the Unmanifest to flow to the manifest. When the packet of energy [violence-Unmanifest-violence] reaches a critical level, it explodes [information transfer] in the realm outside space-time and this is governed by superluminal channel. It travels deep in water and gets locked up in some "*fault*" in the ocean bed. The fault with locked up energy explodes in time as volcano and gives rise to tsunami.

When a conditioned mind-brain-body system throws such a packet of energy in the atmosphere, it is stored up in the atmosphere and from there it goes deep down the earth and gets stored there. Physics tells us that energy cannot be destroyed, so it is locked up deep in the earth, frozen in time. If the stored up energy cannot be sustained, it gets unlocked and gets dissipated as a earthquake or volcano eruption. When the energy of violence attains a critical mass, it explodes as a earthquake, but this criticality is not amenable to any measurement. If there is any 'fault' in the earth's crust it appears as earthquake which has an elemental dynamics of its own. The earthquake is defined as an explosion when the energy of violence has a critical mass and coupled with the elemental dynamics of the fault in the sea bed.

The packet of violence energy is locked up deep in the ocean frozen in time. In a suitable moment [governed by Unmanifest] it unlocks, acts as a earthquake, and gives rise to tsunami.

Fritjof Capra, an eminent physicist, discussed 'natural' disasters in 1998 that are caused in part by human-induced global climate change. He stated that Hurricane Mitch took 10,000 lives, the catastrophic flood of the Yangtze River in China caused more than 4,000 deaths, and Bangladesh suffered floods killing 1,400 people. Capra added that "the rising tide of natural catastrophes over the past decade is a clear indication that the climatic instability caused by human action is increasing".[7] Janet Abramovitz, World Watch Institute, wrote, "The 'unnatural disasters' were made more frequent or more severe due to human action".[8] Capra concluded that, "the principal lesson to be learned is that the cause of most of our environmental and social problems are deeply embedded in our economic systems".

The recent report of the Intergovernmental Panel on Climate Change [IPCC], the United Nations today talked of Greenhouse Gases [GHG] such as carbon dioxide, methane and nitrous oxide released by human activities are warming the world abnormally. The IPCC, drawing on round 2,500 scientists worldwide has confirmed in stark figures. This poses a critical security risks. Global climate changes pose significant risks to the planet. This is "very likely" caused by human activity. All disasters are constructs of the man-made cosmos. Obviously the answer is embedded in a higher level of consciousness of the system.

The art of living is a harmonious movement of understanding, learning, acting – this bestows direct perception or communion with nature. Meditation is not a state; it is dynamic movement of unconditioned energy without any conflict or observer, or meditator.

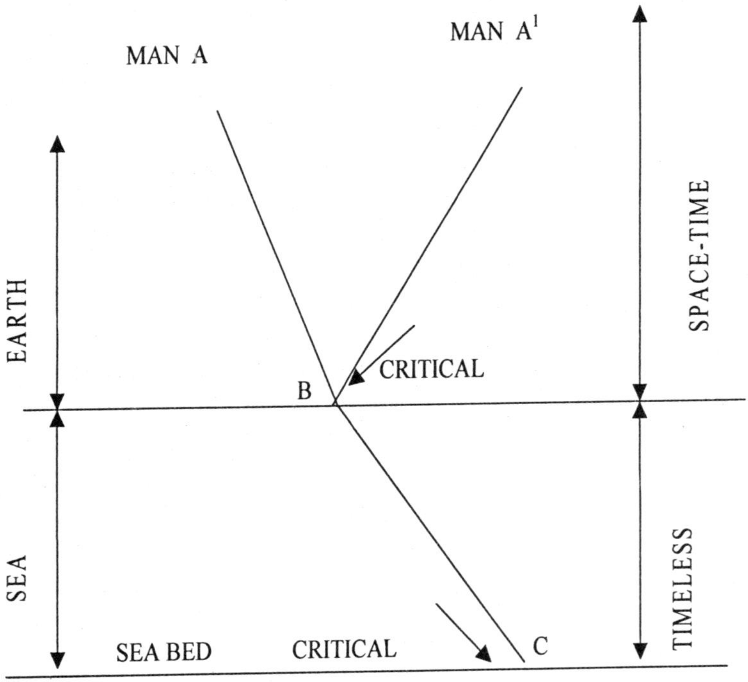

Fig. 1. An artist's vision: (i) the violence packet travels from A to B and from A[1] to B in time-space, (ii) it becomes critical at B and it travels from B to C in the timeless realm; (iii) it gets frozen in time at C; (iv) it expresses itself in time which expresses as earthquake; and (v) the earthquake transforms itself as tsunamis.

An exploration indicates that : (a) the truth of man and truth of nature emerge from the Cosmic Consciousness or Life Field; (b) the catastrophes are constructs of the conditioned psyche over a long chronological time; and (c) this suggests a new relation between earthquakes/tsunamis and man and nature. It is basically an intuitive insight and rejects intellectual approach. Fundamental discoveries which go deep into the nature of things are the result of uncommon intuition and faith, not cold logic. Intuition is fundamental and intellect is a derivative. It shows the unity of Life Field, or Oneness of the Cosmic Consciousness.

An artist's vision in shown in Fig. 1. This model is supported by eminent scientists - say Capra, World Watch Institute and IPCC, and by eminent sages and saint.

To sum up, all droughts, floods, catastrophes, calamities, earthquakes, tsunamis, snow & avalanches, Global Warming are not due to nature's wrath against mankind, but are constructs of the conditioned mind-brain-body system or inheritance of man over uncountable aeons. An understanding gives you the feeling that man divorced from nature could not be happy and man's power to resist nature is hopelessly puny.

3. Global Warming

The IPCC discussed global warming which is becoming an internationally-recognised reality. The gradual melting of Himalayan glaciers and decreasing snow-fall in recent years have contributed. It is an unprecedented crisis. As per media reports, "the government is planning to set up a regional center for monitoring regional environment and climate change".

There must be fundamental radical revolution. Reform needs further reform and this is an endless process. The only revolution is the revolution between man and man, between man and environments, between man and nature. Actually it is only the action of will [ego, I-ness] and assertiveness that needs to be changed at all, because the only mischief in relationship is conflict, between individuals or within individuals, and conflict is will and assertiveness. Conflict is our main concern. Violence is conflict because it is antagonistic to the immutable laws of nature. We have to end all violence, conflict, contradiction. Obviously, it is due to socio-economic-political problems, which compel us to live in a man-made world.

4. Mahatma Gandhi's Teachings in Response to Bihar Earthquake

A great calamity – known as Bihar Earthquake – which descended on India on January 15, 1934. Nearly 25,000 people died in the earthquake. Many more thousands suffered from injuries sustained and still more homeless.

Mahatma Gandhi, in a prophetic language, wrote the following:

Even in this indescribable calamity there is a divine purpose that works for the good of humanity.[9]

To me the earthquake was no caprice of God nor a result of meeting of mere blind forces .... Visitation like droughts, floods, earthquakes and the like, though they seem to have only physical origins, are, for me, something connected with man's morals.[10]

And since God is love, we can say definitely that even the physical catastrophes that He sends now and then must be a blessing in disguise.

We have to understand, learn and live in harmony with nature and harmony with all human beings.

Sri Aurobindo wrote, "A catastrophe may intervene and interrupt and destroy what is being done, but even then the final result is sure .... A new spirit of oneness will take hold of the human race".[11] The message in 1967 of the Divine Mother [Madame Mirra Alfasa] read, "Men, countries, continents! The choice is imperative: Truth or the abyss". Mother had once observed, "Three-fourth of the humanity is obsolete".[12] All this suggest that (a) since 1947, the conditioning or violence has increased tremendously; (b) man has started living in a conceptual world, a world of traditions; (c) lowering the level of sensitivity; and (d) this is leading to self-destruction and self-annihilation.

Paul Davies, an eminent physicist, proposed a unified theory of all existence. He wrote, "Few physicists have been deeply inspired by the remarkable harmony, order and unity of nature that recent advances have uncovered .... Every advance in fundamental physics seems to uncover yet another facet of order .... There is unexpected harmony and coherence of nature .... We cannot fail to be struck by the ingenuity of the natural world? Nature is astonishingly clever in the way it gets thing to work .... The laws which enable the universe to come into being spontaneously seem themselves to be the product of exceedingly ingenious design .... The universe must have a purpose".[13]

5. Snow & Avalanches

It is possible that packets of energy of violence [violence-Unmanifest-violence] may be locked as frozen in time in some hill. If this energy reaches a critical level, there is some relative movement of different positions of the hill. This relative movement leads to snow avalanches.

India being a nation where natural calamities strike every now and then, there is the need for trained disaster management professionals is felt acutely. Further as natural disasters are unpredictable and unknown, we have to design the fore-warning instruments.

6. Underground Observatories

The Sudbury Neutrino Observatory near Ontario results have provided revolutionary insight into the properties of neutrinos and the core of the sun. SNO is a heavy-water built 6800 feet [2000 meters] under ground. Art McDonald says, "There are changes in the neutrinos themselves as they travel from core of the Sun to the Earth." The nuclear reactions fuel the Sun. Every second, 60 thousand million neutrinos from the Sun pass through each square centimeter of the Earth's surface and through you !. "The determination that the electron neutrinos from the Sun transform into neutrinos of another type is very important for a full understanding of the Universe at the most microscopic level.... The direct evidence for solar neutrino transformation also indicates that neutrinos have mass .... The neutrinos make up only a small fraction of the total mass and energy extent of the Universe" In other words, "neutrinos have mass and they change, or oscillate, into another form enroute from the Sun to the Earth"[14]

The scientists of the SNO wrote "The neutrinos has proved incredibly elusive. I am thrilled at the results... The scientists have recorded neutrinos [about 10 per day]... The engineering requirements alone are mind boggling." It is surprising that the violence-conflict energy packet is many thousand times on the surface of earth, so it does result in huge loss of life and property. It is incredibly exciting that scientists have done nothing regarding man-made world of violence and strife which has given birth to such global horrors as tsunami, earthquakes and snow & avalanches, and so on.

We cannot rule out packets of "violence-Unmanifest-violence" energy traveling deep down underground. Even if we reject all insights of religious men, we cannot explain how harmonious, ordered and benevolent nature acts as a tyrant in case of earthquakes/tsunamis. This exploration is only possible by a scientist who lives in a non-dual and non-conceptual frame of reference of Mind Continuum – we need intuitive wisdom.

7. New Relationships

The new relationship of the great calamities, including earthquake and tsunami, is that instead to rupture and opposition between man and nature, there is a growing coherence of our insight [intuition] between man and nature. The aim of the physical calamity is to remind the human race that it must jettison the antagonistic attitude and learn to live in "harmony and order", "non-violence and peace", and "love and compassion" dimensions of life. Man has to understand, learn and live in equanimous awareness.

Every event/organism is governed by the immutable laws of Life, by cosmic Consciousness, by Life Field. Everything is prearranged and everything is an effect. ***The truth of man and the truth of nature is the 'Oneness of Life'.*** Every expression of nature is interwoven and interlinked with the true nature of man. Man is the miniaturised cosmos[15].

8. Freedom

Albert Einstein wrote, "I do not believe we can have any freedom at all in the philosophical sense for we act not only under external compulsion but also by necessity".[16] Schopenhauer's saying, a man surely do what he wills to do, but he cannot determine what he wills".[17] The psychological mode is a kind of prison for us".

Nature, out of benediction, has bestowed on us a hidden potential – *intuition, invention* and *discovery*. This bestows us a freedom and a foundation for inner security. This dimension is free from all violence, conflict and contradiction. If man lives in this dimension, all disaster, including tsunami, earthquake, calamities will go to abeyance.

The task of turning the age-old tide of human ambition and aggression is formidable. The large gap between the superhuman task [freedom] and the frailties of the human endeavors [education of the mind-brain-body] are colossal. Max Born, at the third Pugwash Conference, stated, "clever, rational ways of thinking are not enough. The danger of mass slaughter .... be overcome by moral conviction, by determination to replace national prides and prejudices with human love".[18]

9. Violence and Conflict

Albert Einstein stated, “It is in fact, nothing short of a miracle that the modern methods of instruction have not yet entirely strangled the holy curiosity of inquiry, for this dedicate plant, aside from stimulation, stands mainly in need of freedom, without this it goes to wreck and ruin without fail”.[19]

He, out of his later years, wrote, “nothing of all that will remain but a few pages in the history books, briefly picturing to the youth of future generation the follies of its ancestors.[20] Pupul Jayakar wrote, “A general destructive violence, a fragmentation at all levels, and a growing lack of sensitivity was permeating the landscape of India”.[21] J. Krishnamurti said, “He noticed the general decline in values and the tendency to think responsibility [This] is happening inwardly. One sees .... a growing indifference to man and environments”.[22]

During the last five decades, the terrorism, hijacking, religious fundamentalism, ethnic conflicts, organized crime, drug trafficking, corruption, pollution, population transfer are different facts of terrorism. These are constructs of the conditioned mind-brain-body system, the polluted psyche.[23]

Man cannot escape his responsibility. If he avoids “educating the mind-brain-body system”, it leads to self-annihilation. The disasters are bound to be more frequent and more severe due to human action”. This is bound to add to human suffering and pain.

Nicholas Stern wrote, “It would be for cheaper to take strong measures now to cut emission and stablise greenhouse gas concentration at acceptable levels than to face the economic consequences of catastrophic climate change [24]. In a recent meeting in Washington D.C., legislators gave that The “GS +5” account for two-thirds of global greenhouse gas emission. The problem of limiting climate change is a serious one.

10. Ending of "Man-Made" Disasters

We have to restructure our science – *"Science-Education-Ethics-Health"*. We need to restructure society which is not based on symbolic-dualistic thought, but based on non-dual and non-conceptual thought. This also leads to arresting the fall of civilisation and flows to rise of civilisation. We have to understand and live according to the immutable laws of nature. It implies that there is an urgency to understand the "Symbiosis of Science and Spirituality".

Paul Davies, an eminent physicist wrote, "The laws which enable the universe to come into being spontaneously seem themselves to be the product of exceedingly ingenious design .... The universe must have a purpose, and the evidence of modern physics suggests strongly to me that the purpose includes us".[25]

The non-dual frame is invariably harmonious and ordered, there is direct perception and it is in communion with the laws of nature. The trans-psychological revolution is free of all chaos and conflict, and nature's every expression is in harmony, order and coherence.

The concepts are grafted on the perception, which reduces a perception to a concept. We are not aware of direct perception, but are aware of the mutilated perception or conceptual model. This is known as the crises of perception. We have to jettison the concepts and imbibe direct perception.

There must be complete open-ness or receptivity. We have to observe non-subjectively, without any evaluatory comparisons or judgements.

Thought: We call the boundary, or the wall, between the symbolic-dualistic frame of mind spectrum and non-dual and non-conceptual frame of reference of Mind Continuum as the "frequency barrier", or the "threshold frequency barrier", or the "event horizon". And events below the frequency barrier is within the speed of light and any happening above the frequency barrier is beyond the speed of light. The innovation is bound by "speed of light", while the mutation is *"beyond the speed of light"*. Thought can never penetrate the problem of human relationship. Thought is superficial and old and is the out come of the past. The past cannot enter into something that is totally new. It means that psychologically the thought is sterile in bringing any mutation. The observing and understanding is in the non-dual frame of Mind Continuum, so understanding is not an intellectual affair.

A quiet mind is exceedingly sensitive and aware and spontaneously receptive. A quiet mind is at a higher quantum energy potential of consciousness than a chattering mind.

Man has to educate his mind-brain-body system, and this education will flow to ending of all violence and conflict. It is the higher expression of Consciousness that can heal the conditioned psyche. When the Mind revels in Intelligence, all man-made disasters vanish. It is imperative that "the Union of Concerned Scientists [UCS]" and "The United Nations" issue a manifesto for education of Mind. There is an urgency to initiate an open dialogue with freedom and non-authoritarian way between scientists-philosophers-mystics.

The essence of mind's research (called spirituality) has been highlighted by Andre Malraux in words, "The third millennium will be spiritual or there will be no third millennium"[25] Rudolf Schneider wrote, "A new spiritual pyramid must be erected on the high intellectual platform ...."[26] The declaration of the Sacred Earth gathering Rio-92 wrote, "The World Community must act speedily with vision and resolution to preserve the Earth, Nature and humanity from disaster. The time to act is now. Now or never.[27]

The whole cosmos is pulsating with an eternal, infinite, immutable, beginningless and endless Universal

Consciousness, the Unmanifest, the highest frequency Conscious Energy. Every existence (physical, biological and psychological) – atom, event, happening – is a cosmic dance of the *play* of the manifest and Unmanifest, limited and Limitless, measurable and Immeasurable. Only the *Homo sapiens* – man – is endowed with the capacity to perceive the Unmanifest, the highest expression of Consciousness; and to live in the light of the Divine Energy. This is the only vocation of man. This is the consummation of human life.

11. Concluding Comment

All droughts, floods, catastrophes, calamities, earthquakes, tsunamis, snow & avalanche are not due to nature's wrath against mankind, but are constructs of the conditioned mind-brain-body system or inheritance of man over uncountable aeons. It reminds man that the age-old tide of human ambitions and aggressions must be jettisoned. And naturally can be ended when the human mind-brain-body is purified.

We have to restructure science so that it takes in its realm both the objective and the subjective dimensions of life. The restructured science has the dimension:

[Science – Education – Ethics – Health]

The ethics imply to take in its ambient the sensual and trans-sensual dimension of life. A living in the *present,* the *now* heals and living in the *past* and *duality* suppresses man's perspective sensitivity and enhances disease. We must be aware of the mutation of the whole. It is the higher level of awareness that heals.

We have to jettison human intellect and reason, its information [knowledge] and experience and imbibe intuition. The *intuition, invention* and *discovery.* belong to the expressions of nature. Hideki Yukawa quoted Galileo Galilei in words, "He asked questions addressed to nature and succeeded in receiving.

answers directly from nature.[28] Herman Weyl wrote, "I have a direct knowledge entirely different from the theoretical knowledge that represents the "parallel" cerebral processes in symbols. This inner awareness of myself is the basis of the understanding of my fellowmen".[29] And Ilya Prigogine and Isabelle Stengers completely agree with this.[30]

There is nothing in this universe that belongs to man. Voltaire prophetically expressed the truth that, ".... everything is governed by immutable laws .... everything is prearranged .... everything is a necessary effect .... we are both equally foolish, both are toys in the hands of destiny"[31] .

In Love, or Truth, or Beauty, or Reality, there is superluminal [faster-than-light transfer of negentropy.[32] In nature everything is a happening.

In nature everything happens simultaneously and everything is interrelated. It is integrated. It is self-emerging system, a totality that exists. Life is mysterious wholeness. It is non-fragmentary, indivisible whole. It is homogeneous, undifferentiated and dimensionless. It is timeless, eternal and infinite. Life is immeasurable and unnamable.

The creativity [*intuition, invention* and *discovery*] are nature's direct perception; and this leads mankind out of the prison of his own conditioned mind. Nature, in its benediction, is full of harmony and order, non-violence and non-conflict, love and compassion, perfect sustainable freedom and peace, and free of all psychological imbalances and contradictions. The ending of man-made disasters happens when man engages himself in "objective observing and understanding', or "self-direct-perception", or holy curiosity of inquiry. One is deeply struck by his concern about the large gap between the superhuman task [of transformation or mutation] and the frailties of human endeavour"[33]

Krishnamurti wrote, "something has been destroyed in India, but a presence waited in the land. The benediction of the timeless is waiting, the ground is ready. Can we create something that is of this benediction"[34]. It was a prophecy that a sacred energy was alive again in the Indian soil.

If the conditioned mind voluntarily goes to non-action, the Cosmic Intelligence takes over the change of our being, and all disasters and catastrophes go to abeyance, and man lives in ***"sat-chit-ananda"***, or love-freedom and peace.

References

1. Albert Einstein, *Reader's Digest,* July 1973, p. 133.
2. The Tribune, May 20, 2005; *Monster Explosion in Milky Way Spotted.*
3. The Union of concerned Scientists, *World Scientists' Warning to Humanity,* USA: April 1993.
4. Pratibha Chauhan, *The Sunday Tribune,* Chandigarh, February 4, 2007; and The Hindu, February 10, 2007 [editorial].
5. Sampooran Singh and Dr. (Mrs.) Kanwaljit Kaur, *Symbiosis of Science and Spirituality,* Delhi: Kelpaz Publications, 2006, pp. 345-380.
6. St. Petersburg, *UNI, PTI, The Tribune,* Chandigarh, May 25, 2005.
7. Fritjof Capra, *The Hidden Connections,* New Delhi: Harper Collins Publishers, India, 2003, pp. 183-184.
8. Janet Abramovitz, in Fritjof Capra, *ibid.,* p. 184.
9. *Harijan,* Feb. 2, 1934; *The Hindu,* Jan. 24, 1934 and *The Hindustan Times,* Jan 25, 1934.
10. *Harijan,* Feb. 2, 1934; *The Hindu,* January 26, 1934; and Feb. 16, 1934.
11. Sri Aurobindo, *A Message on August 15, 1947;* in The Tribune, August 15, 1997.
12. The Divine Mother [Madame Mirra Alfasa], in *CIR HU, Centre of International Research in Human Unity,* Auroville, Paper 4, Feb. 1997, p. 55.
13. Paul Davies, *Superforce,* New York, Simon and Schuster, 1984, pp. 224; in Sampooran Singh et. al., p. 370.

14. Sudbury, Neutrino [SNO] Observatory.http://www.sno.phy.queensu.ca.
15. New Relationships, in *Symbiosis'of Science and Spirituality, op. cit.,* p. 379.
16. Albert Einstein, *I Believe; Nineteen Personnal Philosophies,* Unwin Books, London, 1962, p. 26.
17. Schopenhauer, in *Albert Einstein,,* ibid, p. 26.
18. Max Born, in Hideki Yukawa, *Creativity and Intuition*, Tokyo : Kodansha International Ltd, 1973 pp 189-190.
19. Albert Einstein, in *Nature, Vol. 278*, 15 Mar 1979, p. 216.
20. *Ibid, Qut of My Later Years*, Greenwood Press, Connectual Ed. 1970, p. 8.
21. Pupul Jayakar, *Krishnamurti : A Biography*, San Francisco : Harper & Row Publishers, 1986, p. 314.
22. J Krishnamurti, in Pupul Jayakar, ibid. 314.
23. Sampooran Singh, et.al., The *Masterplan, Paradigm for Human Survival and Excellence,* Chandigarh, Faith Publishers, February 1998, pp 126 - 168.
24. Nicolas Stern, in The Hindu, Feburary 23, 2007, Editorial.
25. Andre Malraux, in Robert Muller, *My Testament to the UN,* 1993, Ch. 9.
26. Rudolf Schneider, in *Plenetary Synthesis,* in Robert Muller, *ibid*
27. Declaration of the Second Earth Gathering, Rio-92 in Robert Muller, *ibid.*
28. Galileo Galilei, in Hideki Yukawa, *Creativity and Intuition,* Tokyo: *op cit.,p. 111.*
29. Herman Weyl, *Philosophy of Mathematics and Natural Sciences,* Princeton, N.J. : Princeton University Press, p. 311.
30. Ilya Prigogine and Isabelle Stengers, *Order Out of Chaos,* Toronto: Bantam Books, 1984, p. 311.
31. Voltaire, *Dictionnaire Philosophique,* Paris: Garnier, 1954, in Ilya Prigogine, et. al., 1984, pp. 257-258.
32. Life: Superluminal Transfer of Negentropy, in *Symbiosis of Science and Spirituality, op. cit.,* p. 377.

33. Victor F. Weisskopf in Hideki Yukawa.op.at., Indruction p. 13.
34. Pupul Jaykar, in *Krihnamurti: A Biography, op.cit* p. 380.

5

The Immanent Disaster

1. Climate Conference – 1979-1988

In 1979, the first ***"World Climate Conference"*** organised by the World Meteorological Organisation [WMO] expressed concern that "continued expansion of man's activities on earth may cause significant extended regional and even global changes of climate". There were many subsequent conferences – say 1985, 1987. At its 40th Session in 1988, the WMO Executive Council decided on the establishment of the Intergovernmental Panel on Climate Change [IPCC] in collaboration with United Nations Environmental Programme [UNEP]. In November 1988, the UN: IPCC held its first Plenary Session.

2. The Aims and Objects of UN: IPCC

The UN: IPCC met between December 3 and 14, 2007 in Bali. About 192 countries took part in this conference. The IPCC had unequivocally affirmed the warming of our climate system and linked it to human activity.[1] The UN Secretary General Ban Ki-moon stated that the immanent disaster [global warming] facing us is "even more terrifying, because they are real". Hans Verolme wrote, "We have caused climate change and it is also clear that we hold the solution to stop global warming in our hands".

The UN: IPCC stated that we need to limit concentration of CO_2 to 400 ppm [parts per million of CO_2 equivalent] or below [compared with 280 ppm in the pre-industrial era and 380 ppm today].[2] We have been through the Clean Development Mechanism [CDM] for about two decades, but the progress in minimal. Obviously, the CDM is inadequate to resolve our challenges.

The role of the IPCC is to explore the wholeness of the challenge – objective and subjective, open and transparent – to understanding the *scientific analysis* of the "human-induced climate change", its potential impacts and options for adaptation and mitigation".[3]

3. The Immutable Laws of Nature

Observing [seeing], listening [hearing], touching, taste is direct; thinking, thought, inheritance is indirect. Voltaire wrote that ".... everything in nature is governed by immutable laws .... everything is prearranged .... everything is a necessary effect".[4] All activities of Nature are interrelated and interpenetrated, and are in harmony, order and peace.

Hideki Yukawa, a Japanese Nobel laureate, wrote, "Science is a manifestation of man's efforts to open up the unknown that lies before him, to discover new possibilities". There are two methods of thinking – one is 'induction' [direct, fundamental, ***mystical intuition***] and the other is 'deduction' [indirect, derivative, ***scientific analysis***].[5] All expressions of Nature are direct.

Albert Einstein, the greatest physicist of the age, discussed "the nature of God, the universe and man" and said, "We know nothing about it [how the violence-conditioned-patterns change to earthquakes, tsunamis, and global warming] at all. Our knowledge is but the knowledge of schoolchildren .... The real nature of things – that we shall never know, never".[6]

In the third Pugwash Conference, Max Born, an eminent scientist, wrote, "Clever, rational ways of our thinking is not enough. The danger of mass slaughter .... can only be overcome by moral conviction, by a determination to replace national prides and prejudices with human love".[7] It is the Love or Truth, that can resolve all human conditionings. Love is that state in which the thought process, as time, has completely ceased.

4. Civilization is a Process of Conditioning

Man had an evolutionary origin at about 2.5 million years ago. The thought's origin is embedded in deep antiquity and slowly it developed to reason. We have travelled through millions of years, through centuries with these words, symbols, ideas and concepts.

The mind transferred the physical to psychological world. What we call mind is a conditioned energy moving through the body, monitored by brain, operating through the senses. Mind is consciousness and consciousness is an energy.

Psychological time-space, or I-consciousness, or me, or ego do not have any factual reality, or objective reality. These are just concepts, constructs of the relative frame of reference of the mind spectrum. The ego has a conceptual reality. It has an ideational reality.

Man for uncountable aeons has been experimenting to end ego [transcend ego], but till todate he has totally failed. Perhaps we have to depend on inductive logic. The hypothesis is supported by eminent scientists – say Albert Einstein – and eminent saints/sages – say Mahatma Gandhi, Sri Aurobindo, Vimala Thakar, Lama Govinda, John Selby, Eckhart Tolle, Lord Buddha.[8] Every word uttered by these saints/sages is in the timeless realm. Nature's every expression is in the timeless realm and free from any violence, conflict, contradiction, and imbibe freedom and peace.

The "human-induced climate change", or "human activity" implies that these are caused by conditioned mind-brain-body system, or by the symbolic-dualistic frame of reference of mind spectrum. Modern science has shown all the concepts we use to describe nature are limited and creations of the conditioned mind; parts of the map, not of the territory. Every activity of the mind adds to conditionings and enhances the immanent disaster.

Intellectual understanding is not sufficient. Knowledge cannot transform the inner core of our being. Knowledge is sterile. Knowledge will enhance calamities/catastrophes/disasters/earthquakes/tsunamis/global warming.

Violence: Civilization is a process of conditioning; the ego adds to conditionings; it is identification with ideas or concepts. H. K. Dua wrote, "India is increasingly becoming a violent nation .... And those running the administration are unable to see that unchecked violence can create more tensions, sharpen divides, cause political instability and unsettle the democratic system".[9]

All psychological imbalances – violence, fear, greed, anger, and so on – are constructs of the waking state of mind, or relative frame of reference of mind in the mind spectrum. It is an indication of psychological immaturity. Non-violence is a percept in the non-dual and non-conceptual frame of reference of the Mind Continuum. You cannot impose non-violence on a consciousness which is violent because violence has deep rooted causes in the psyche. A mutation is possible in the brain through insight, not through knowledge, not through accumulation. That insight is the ability to see a truth directly, a direct perception of the truth.

5. New Relation between Immanent Disaster and Man and Nature

The technology of 'induction' in science was advocated by Francis Bacon and effectively used by Galileo Galilei. It starts with comparison of more or less similar experiences and experimental results. It is an introspection by eminent scientists/saints/sages, which is expressed in the Eternal Present or Timeless Realm.

An exploration indicates that: (a) the truth of man and truth of nature emerge from Cosmic Consciousness or life field; (b) the calamities are constructs of the conditioned psyche over a long chronological time; and this suggests a new relation between calamity and man and nature. It is basically our intuitive insight and rejects intellectual approach. Fundamental discoveries which go deep into the nature of things are the result of uncommon intuition and faith, not cold logic. Intuition is fundamental and intellect is a derivative. It shows the unity of life field, or Oneness of the Comic consciousness.

In the symbolic-dualistic thought, the observer-knowledge-experience-memory gives birth to violence in the internal realm and this is called direct. This violent-thought travels from the inner realm to outside field as violence-Ummanifest-violence and this is called derivative in the external field, which is time till it reaches the earth or the sea bed. Osho wrote, "Between two thoughts there is an interval – in the interval there is consciousness .... Pay more attention to the background [Consciousness] and less attention to figure [violence]".[10]

(a) Bihar Earthquake: A great calamity descended an India on January 15, 1934. Nearly 25,000 people died in the earthquake. Many more thousands suffered from injuries sustained and still more homeless. Mahatma Gandhi, in a prophetic language, wrote the following:

Even in this indescribable calamity there is a divine purpose that works for the good of humanity.[11] .... To me the earthquake was no caprice of God nor a result of meeting of mere blind forces .... Visitation like droughts, floods, earthquakes and the like, though they seem to have only physical origins, are, for me, something connected with man's morals.[12] .... And since God is love, we can say definitely that even the physical catastrophes that He sends now and then must be a blessing in disguise.

We have to understand, learn and live in harmony with nature and harmony with all human beings.

(b) Disasters in 1998: Fritjof Capra, an eminent physicist, discussed 'natural' disasters in 1998 that are caused in part by human-induced global climate change. He stated that Hurricane Mitch took 10,000 lives, the catastrophic flood of the Yangtze River in China caused more than 4,000 deaths, and Bangladesh suffered floods killing 1,400 people. Capra added that "the rising tide of natural catastrophes over the past decade is a clear indication that the climatic instability caused by human action is increasing".[13] Janet Abramovitz, World Watch Institute, wrote, "The 'unnatural disasters' were made more frequent or more severe due to human action".[14] Capra concluded that, "the principal lesson to be learned is that the cause of most of our environmental and social problems are deeply embedded in our economic systems".

(c) Paul Davies, an eminent physicist, proposed a unified theory of all existence. He wrote, "Few physicists have been deeply inspired by the remarkable harmony, order and unity of nature that recent advances have uncovered .... Every advance in fundamental physics seems to uncover yet another facet of order .... There is unexpected harmony and coherence of nature .... We cannot fail to be struck by the ingenuity of the natural world? Nature is astonishingly clever in the way it gets things to work .... The laws which enable the universe to come into being spontaneously seem themselves to be the product of exceedingly ingenious design .... The universe must have a purpose".[15]

(d) The new relationship of the great calamities, including earthquake, tsunami, global warming is that instead to rupture and opposition between man and nature, there is a growing coherence of our insight [intuition] between man and nature. The aim of the physical calamity is to remind the human race that it must jettison the antagonistic attitude and learn to live in "harmony and order", "non-violence and peace", and "love and compassion" dimensions of life. Man has to understand, learn and live in equanimous awareness.[16]

6. Psychodynamics of Mind-Brain-Body System

(a) Freedom: It is pure observation without direction, without fear of punishment and reward. Freedom is without motive; freedom is not at the end of evolution of man but lies in his first step of his existence. In observation one begins to discover the lack of freedom. Freedom is found in the choice of awareness, in the attentiveness of our daily existence and activity. Freedom comes, when even momentarily, thought or pseudo-subject goes to abeyance and attentiveness takes over as the true-subject. Pure observation bestows insight, which is without any shadow of past or time. The timeless insight brings about a deep radical mutation in the conditioned psyche or mind-brain-body system.

Understanding the nature of bondage is the beginning of freedom. Understanding the content of bondage gets converted into factual urge for freedom. Freedom is spontaneous equibalance or equipoise of mind. The human being seems to be born with the faculty of self-awareness and understanding, bondage and freedom, dualism and non-dualism.

The aspiration of unconditional freedom in every individual is born when one transcends the dualism of mind-brain-body system which flows into dropping the unreal and the process of becoming comes to an end. This results in a harmonious, non-violent and peaceful life.

Einstein stated "I do not believe we can have any freedom at all in the philosophyical sense, for we act not only under external compulsion but also by inner necessity".[17] Amartya Sen wrote, "Development in its true sense, is an enabler of human freedoms and well-being, rather than a enhancement of inanimate objects of convenience. It is inseparable from environmental and ecological concerns such as clean air and water, epidemic-free surroundings and the preservation of all life forms,".[18]

(b) Peace: It is a total way of living. As long as there is inherent violence in the human psyche, there can never be sustainable peace in the world. Peace requires the mutation of individual psyche. It is only holistic education that helps man to grow from knowledge to understanding, from confrontation to cooperation, from rationality to intuition, from intellect to intelligence.

We are on the crossroads of destiny; we must take the road from information gathering and experience to understanding and learning, from violence to non-violence, to motivation of sharing rather than to dominating, complementarity and not competition, harmony and not discord; and in the final analysis education alone that can bring this about.

Peace is a state, when thought-noise or thought-time ends in that frame of reference of the Mind Continuum. A sustainable peace is defined as when the mind attains the capacity to remain in the state of attention or a state of awareness. Peace requires free human beings. The search for peace is a search of new dimension of consciousness.

We want a psychological revolution, in which the whole human race has to participate and every man is an equal partner. This collective psychological experiment will awaken the integral and all-inclusive dimension of Consciousness and that can guarantee human survival along with all fauna and flora.

(c) Meditation: It is a way of living where relationships are movements of relaxation and not tensions and pressures. When the identification with functional roles ends, there is relaxation of the mind. When the mind relaxes into non-action, the mind-brain system is free of tensions and pressures. Meditation bestows silence and equipoised mind, so silence is a tension and pressure-free state of your being. The state of total relaxation – physical relaxation and mental relaxation – is a marvelous state. Silence implies that you are aware with the wholeness of your being. From the partial and the fragmentary view of life one has moved into the wholeness. And it is the wholeness that heals. Meditation is a sacred journey to the ***Wholeness of Life.***

(d) The Holistic Education: It must educate the psychological structure and physical organism to move into non-action and non-dual frame of reference of Mind Continuum.

Education implies to find out the value structure at every moment. Understanding is for living. If you understand the truth that sustains the value structure, the society gets transformed at that moment. The understanding of truth that governs the value structure constructs a sane global society. All spiritual values arise from the realm of the Infinite, the Eternal. ***Human life is an opportunity to manifest the harmony, the order and the Intelligence that we observe around us in cosmic nature.***

(e) It implies that when a lower frame of reference comes in contact with non-reactional attentivity, it makes a quantum jump to a higher quantum energy potential of Consciousness; and it is the consciousness that resolves all challenges.

(f) Victor F. Weisskopf wrote, "Man's power to resist nature is hopelessly puny …. How difficult is the task of turning the age-old tide of human ambition and aggression …. There is a large gap between the superhuman task and the frailties of human endeavors".[19] A word of caution, man has been trying to end ego or I-consciousness for uncountable aeons, but has utterly failed so far.

(g) I humbly suggest that the IPCC may examine to implement both the fundamental *[mystical intuition]* and the derived programmes *[scientific analysis]* as both are essential for human survival; and we have to work for low-carbon technologies. We call upon scientists in the world to join us in our survival and excellence.

7. Resolution of Human Paradox

(a) Albert Einstein, in prophetic words, said that we need, "Objective observing and understanding" to resolve our psychological imbalances; and this leads to "peace end serenity".[20] Thought comes in a symbolic-dualistic frame of reference of mind spectrum. As soon as thought comes, the naming, the comparison, the evaluation comes up involuntarily. The thought gets stored in the memory and this is called the conditionings, or psychological patterns. These are expressed as nervous tensions or chemical pressures.

(b) In the symbolic-dualistic frame of reference of mind spectrum, the ego and the thought are coupled. These are inseparable. The ego identifies with various roles. If an inner urge is born to observe, the conditioned mind makes a quantum jump to a higher state of consciousness – non-evaluatory perception. The symbolic-dualistic frame of reference makes a quantum jump to non-dual and non-conceptual frame of reference of Mind Continuum. In other words, in the non-dual and non-conceptual frame, the ego drops itself so the movement of thought-structure is not a voluntary movement at all, it has nothing to do with his will, noting to do with his desire, it has its own independent movement. In other words, the movement of thought-structure is independent of your volition. Observation of thought implies not to judge, neither to accept nor to reject, not to act upon them, not to change them. I look at the movement of thoughts in utter freedom. The mental movement and a momentum is independent of "me". There is nothing like an individual mind, and there is nothing like an individual "I consciousness". There is racial consciousness and a racial momentum of knowledge-experience-thought; and this gives a great freedom within.

(c) If observation of thought is continued for a long chronological time, both the thought and the thinker coalesce and vanish, and we attain the state of Intelligence, pure Consciousness, which resolves all human challenges.

(d) The life consists of movement and non-movement, motion and non-motion, action and non-action. These two together make the wholeness of life. there is sheer motionlessness. No thought moves, no feeling, no sentiment, no emotion is moving. A movement of thought is equal to tension in your whole nervous system; and any movement of emotions and sentiments is equal to chemical turbulence in the body.

(e) Meditation is a way of living where the relationships are movements of relaxation and not tensions and pressures. Silence is a tension-free and chemical pressure-free state of your being.

(f) Human beings are born for enjoying unconditional inner freedom. It is only free individuals who can cooperate with one another. Science and spirituality follow the law of non-authoritarian approach – an approach not conditioned by the authority of the past.

(g) Education is liquidating the imbalances and stimulating spontaneous equipoise and equibalance. Health is a spontaneous harmony of autonomously working organs in your body. The heart is perfect and whole by itself. In education and in health there is wholeness moving and operating. Silence is the termination of enquiry as a cerebral movement. Silence brings you to the timeless present. Eternity is born every moment and dies every moment.

(h) Life is one whole, it is indivisible. It is homogenous, undifferentiated and dimensionless. In fact, the undifferentiated continuum is dimensionless. Life is an organic wholeness having the divinity of a magnificent interrelatedness. Perception based on division of life into the "me" and the "not me" disappears. The awareness of the interrelatedness awakens Love and Compassion.

(i) Observation Flowering to Intelligence: The occurrence of seeing or hearing, smelling or breathing is a movement of in the sphere of non-duality; and there is no choice in these movements. If this state is sustained for a long chronological time a new energy is born in the state of Silence, and this is called the dimension of Awareness. The energy is released by the non-action of knowledge, and it operates without choice. It is an energy that permeates the whole being; and becomes the source of perception and response and has no relation with the willing, the volition. Awareness or Intelligence is choiceless.

(j) That tension-free, pressure-free, that division-free state, that duality-free state, activises the energies of the Wholeness. Intelligence is the energy of the wholeness. Intelligence is the energy of that complex inter-relatedness of Life.

(k) Understanding and Learning: Understanding happens when the mind is totally quiet and has transcended the time-space matrices, or thought structures. Understanding comes when there is encounter with the facts. Understanding flows to learning. Learning is timeless and every step is the destiny. Learning is a movement of your whole being. It is a dynamic movement of life. there is no psychological time between learning and the being, so it instantaneously becomes the being. This raises the quantum energy potential of the system, so that there is higher understanding. You have to learn non-authoritatively, learning and understandingly.

(l) From Intellect to Intelligence: Man succeeded in discovering reason and developed it through many centuries. The brain is meant for receiving data, processing it, interpreting it, reacting to it. The human brain is a reaction of the past to the present challenge. Mental movement is always dualistic.

Intellect has been misused because it has emphasised only the activity of thought. Intellect, being dualistic, has the capacity to construct a verbal structure as an ideal, a conclusion, something to be achieved. We live in verbal structure.

When the cerebral activity realises its limitations, it makes a quantum jump to higher level of consciousness, and intuition emerges. From the experiencer and the doer, the analyser and the interpreter, one brings oneself into the state of observation, which gives us the intuition. If the state of observation is sustained for a long chronological time, Intelligence is born.

A simple perception can result in an objective approach to all the subjective world; this subjectivification of the subjective world is a tremendous thing. An objective approach to the subjective reactions, subjective feelings wakes up the Intelligence. Every perception and understanding results in growth. The Intelligence is non-personal, and resolves all the psychological imbalances, ends all violence, fear, greed, anger, and so on. It bestows love and compassion.

8. The National Plan of Action for Climate Change

Prime Minister Dr. Manmohan Singh, inaugurated a Council for *"National Plan of Action for Climate Change"* on February 7, 2008. The essence of the plan is: (a) "The Council would release the plan in June 2008; (b) A technology innovations for reduction of energy use by industry and other user sectors; (c) Action by afforestation, drought proofing and flood our river system; (e) We need an international treaty that would provide climatic justice; and (f) "We need friendly technologies". All these are socio-economic-political system essential for the whole world. It is hoped that all these measures may give a limited climate change.

The UN: IPCC must engage itself in a fundamental challenge [*mystical intuition*] and derived challenge [*scientific analysis*]. J. Krishnamurti stated, "It is a crisis in our consciousness". Albert Einstein stated, that mankind needed "objective observing and understanding" to attain "peace and serenity". In other words, we need higher hierarchical level of Awareness/Consciousness to resolve the National Plan of Action for Climate change. The UN: IPCC and other institutions in the world must deal with the Wholeness of life.

The plan of "Climate Change" is a stupendous task. Victor F. Weisskopf, an eminent physicists, wrote, "The large gap between the superhuman task and the frailties of human endeavours" are striking. He added, "Man divorced from nature could not be happy and man's power, to resist nature are hopelessly puny".

(a) Education: No education worthy of the name can be conceived as something external having no relationship "with man's innermost being and becoming", a task of education is inward transformation or inward flowering, and character building. Education is to understand the psychodynamics and limitations of the conditioned psyche and to understand the Unconditioned Energy. The foundation of education is not knowledge [outer] but intuition [inner].

In the words of Ramana Maharishi, "Illiteracy is ignorance and education is learned ignorance. Both are ignorant of the true aim. The true aim is the non-dual mode of knowing which being literate or illiterate concerns the dualistic and symbolic mode of knowing. Ignorance is characterized by non-dual and non-conceptual mode of knowing, which would instantly reveal the universe to the Mind only". Vimala Thakar wrote, "Education if not liquidating the imbalance and stimulating spontaneous equipoise and equibalance"[22]?

J. Krishnamurti said, "War is not in Beirut, it is in our hearts and mind".[23] He added, "Humanity needed to a high level of awareness to solve the problem of nuclear war".[23] We have to undertake the nature and structure of Consciousness. The only panacea for human survival is higher Awareness or Consciousness, so that man perceives the wholeness of life.

(b) A Critique on Creativity and Innovation: Many eminent scientists and philosophers-mystiques have discussed about *'Creativity and Innovation'*. One is creative in the sense that it results in the discovery of truth hitherto unknown to man. Science is a manifestation of man's efforts to open up the unknown world that lies before him, to discover new possibilities. The creativity blosoms to intuition, invention and discovery".[24]

If the scientist engages himself in profit-making, or to make the creativity as a commercial item, he loses the creativity and is caught in a web of 'ego', or 'I-consciousness'. The ego instantaneously brings him at the symbolic-dualistic level – this is called innovation. The "National Plan of Action for Climate Change" basically deals with the socio-economic-political system and this has its own inherent limitations.

The fundamental or direct crisis can solve the challenges, while the derived crisis or indirect crisis deals with deductive logic. In other words, the fundamental crisis is expression of nature, which the derived crisis is indirect or deductive logic, or symbolic-dualistic frame of reference of mind spectrum. The derived crisis constructs "ego" and is bound to cut down the survival period, as the crisis adds to our conditionings.

The fundamental crisis is in our consciousness and we cannot allow it to go on.[25] The crisis can be resolved if we embrace higher level of Perceptive Sensitivity or Awareness or

Consciousness. Where there is love, there is Compassion. And Compassion has its own Intelligence. It is the Intelligence that is the panacea to human survival.

To sum up, man urgently needs "Perception Commission" - and reject "Knowledge Commission" – as knowledge is invariably based on intellect (symbolic-dualistic mind) and "Perception Commission" is based on non-dual and non-conceptual frame of reference of Mind Continuum. The National Plan of Action for Climate Change" must be based on "Perception Commission", as sharing across countries can only happen at Perception or Sensitivity, or Attentivity, or Awareness, or Consciousness.

9. Concluding Comment

(a) All floods, droughts, catastrophes, calamities, earthquakes, tsunamis, snow & avalanches, global warming, are not due to nature's wrath against mankind, but are constructs of the conditioned mind-brain-body system or inheritance of man over uncountable aeons.

(b) Man has to learn to make a quantum jump from intellect to intuition [insight]. It appears that when a packet of energy [violence-Unmanifest-violence] stored in the realm outside space-time and this is governed by superluminal channel; but this release is located in space-time matrices, which appears as earthquakes. All calamities are due to transformation [frozen time] to time.

(c) We unequivocally affirm that objective science alone cannot resolve the "human-induced climate change, and we need *mystical intuition* and *scientific analysis*. We categorically reject the suggestions by eminent scientists that "we hold the solution to stop global warming in our hands".

(d) The IPCC has warned the governments that only a few years left to avert the worst impacts". We present "scientific analysis" that the conditioning of the human psyche is increasing at a rapid rate, and this conditioning is bound to cut down our survival period. It is imperative for every human being, for all political parties, media and governments to cut down all causes leading to rising temperatures, melting glaciers and extreme weather events across the world.

(e) Human-induced climate change adds new stress – as nervous tensions chemical pressures. We want new relations between man and nature, and between man and man. We need a dynamic interplay of objective science and subjective science [spirituality].

(f) The only solution to ending of "human-induced climate change" is higher quantum energy potential of Awareness/ Attentivity/Consciousness. It is imperative to initiate wholistic [holistic] education at all levels at the earliest.

References

1. R. K. Pachauri, *The Sunday Tribune, Nov. 18, 2007 and The Hindu, Nov. 27, 2007.*
2. Sudhir Chella Rajan, *The Hindu,* Dec. 20, 2007.
3. Role of the IPCC, website (www.ipcc.ch) [modified].
4. Voltaire, *Dictionnaire Philosophique,* in Ilya Prigogine and Isabelle Stengers, London: Bantam Books, 1984, p. 257.
5. Hideki Yukawa, *Creativity and Intuition,* Tokyo: Kodansha International Ltd, 1973, pp. 205, 113.
6. Albert Einstein, *The Merry Mind of Albert Einstein,* Reader's Digest, July 1973, p. 133.
7. Max Born, in *Third Pugwash Conference,* in Hideki Yukawa, *op. cit.,* pp. 189-190.
8. Sampoorann Singh and Dr. [Mrs] Kanwaljit Kaur, *Symbiosis of Science and Spirituality,* Delhi: Kalpaz Publications, 2006, pp. 372-374.
9. H. K. Dua, *Editor-in-Chief,* Chandigarh *The Tribune,* Jan. 1, 2008.
10. Osho, *The Book of Nothing,* Pune: Rebel Publishing House, 1975, p. 112.

11. *Harijan,* Feb. 2, 1934; *The Hindu,* Jan. 24, 1934 and *The Hindustan Times,* Jan 25, 1934.
12. *Harijan,* Feb. 2, 1934; *The Hindu,* January 26, 1934; and Feb. 16, 1934.
13. Fritjof Capra, *The Hidden Connections,* New Delhi: Harper Collins Publishers, India, 2003, pp. 183-184.
14. Janet Abramovitz, in *Fritjof Capra, ibid.,* p. 184.
15. Paul Davies, *Superforce,* New York, Simon and Schuster, 1984, pp. 224; in Sampooran Singh *et. al.,* p. 370.
16. Sampooran Singh *et. al., New Relationships, in Symbiosis of Science and Spirituality, op. cit.,* p. 379.
17. Albert Einstein, *I Believe: Nineteen Personal Philosophies,* Universal Books, 1962, p. 26.
18. Amartya Sen, N.L. in *The Hindu,* December 28, 2007.
19. Victor F. Weisskopf, in *Introduction,* Hideki Yukawa, *op. cit.,* pp. 9, 13.
20. Kenneth Brecher, Albert Einstein: 14 March, 1879 – 18 April, 1955, *Nature,* Vol. 278, 15 March 1979, p. 215.
21. Sampooran Singh, in *Vision Tsunamis, Earthquakes, Snow and Avalanches and Global Warming, A Dialogue at Chandigarh,* 02 March, 2007.
22. Ramana Maharishi, in Ken Wibber *The Spectrum of Consciousness*, USA: The Theosophical Publication House, 1982 [Third Quest Book], p. 110.
23. J Krishnamurti, *KFI 1985/3*, pp. 3, 5-6, 19.
24. Hideki Yukawa, *op cit.*, 113, 205.
25. Sampooran Singh, *Physics and Consciousness*, Dagshai: The Atam Science Trust, 2006, Vol. III, pp. 28-41 ibid., A Review on Symbiosis of Science and Spirituality, October 7, 2002.

6

"The Nature of God, the Universe, The Climate Change and Man"

1. A Vision on "Climate Change"

The UN: IPCC Chairman R.K. Pachauri had "Unequivocally confirmed the warming of our climate system and linked it to human activity.[1,2] The objective science reveals the magnitude, geographic extent and precise locations of ocean warming.[3] Humans are reshaping the seas through overfishing, air pollution and commercial shipping. Scientists have analysed factors, including warming ocean temperatures due to greenhouse gas emission, nutrient runoff and fishing.[3] Overall rising ocean temperatures represent the biggest threat to marine systems. Recently scientists presented that the area of Glacier have been reduced from 3,391 to 2,721 sq. km. between 1962-2004. The overall reduction of 21 percent in the glacial surface area was reported. The fragmentation of large glaciers resulting in the mean surface area of glacial extent from 1 sq. km. to 0.32 sq. km. during 1962-2004.[4] However, there is unequivocally evidence of global warming. The retreat parameters of 35 Himalayan glaciers shows that 43 percent of glaciers are receding in the range of 10 to 20 m/year.[5]

One is deeply struck by the superhuman task of resolving the Climate Change and the frailties of human endeavors. This article mainly discusses the human survival. We reassert our conviction that resolution of "Climate Change" or "Global Warming" is realistic and urgent. This work is truly to be seen as a part of a long struggle for the progress of mankind, and scientists have a responsible part.

2. The National Plan of Action for Climate Change

Prime Minister Dr. Manmohan Singh, inaugurated a Council for *"National Plan of Action for Climate Change"* in the Tribune of February 7, 2008. The plan will be released in June later this year. This will help poverty alleviation everywhere. All these reforms are socio-economic-political system essential for the whole world. Reform is an endless process. It is hoped that the plan will give a limited climate change.

J. Krishnamurti gave an answer to this paradox of reforms. He wrote, "The only revolution is the revolution between man and man, between human beings. That is our only concern .... It is only the action of will and assertiveness that needs to be changed at all, because the only mischief in relationship is conflict, between individuals or within individuals and conflict is will and assertiveness".[6] In other words the conflict is ego or I-consciousness, or violence of the symbolic-dualistic frame of reference of mind spectrum.

The UN: IPCC must engage itself in a fundamental challenge *[mystical intuition]* and derived challenge *[scientific analysis]*. It is a crisis in our consciousness. Albert Einstein stated, that mankind needed "objective observing and understanding" to attain "peace and serenity". In other words, we need higher hierarchical level of Awareness/Consciousness to resolve the National Plan of Action for Climate Change. The UN: IPCC and other institutions in the world must deal with the Wholeness of Life.

We have to undertake the nature and structure of Consciousness. The only panacea for human survival is higher Awareness/Consciousness so that man perceives the Wholeness of Life.

3. On Transformation or Mutation or Revolution

Transformation is not in the future, can never be in the future. It can only be *now,* from moment to moment. Truth or Love is a direct [fundamental] expression of the Wholeness of Life. Truth is not cumulative. It is from moment to moment. Eternity is in the moment. Eternity is in the *now.*

Mutation is a radical change in the quality of the mutant. It is not a peripheral change. It is a radical change in the whole of the mutant. Such a mutation is possible in the brain through insight, not through knowledge, not through accumulation. That insight in the ability to see a truth directly, a direct perception of the truth and education must foster that.[7] So we must not only emphasize knowledge, but emphasize insight – the possibility of insight. The human mind must come upon this non-accumulative state – a state of awareness, alertness, watchfulness, observation, learning which is different from the state of gaining, profit, accumulation whether it is knowledge or money or whatever.

You cannot impose non-violence on a consciousness which is violent because that violence has deep-rooted causes in the psyche. Unless these causes are eliminated there is no ending of violence.

George Wald, Nobel Prize in Medicine, 1967, wrote, "That mind, rather than being a very late development in the evolution of living things, restricted to organisms with the most complex nervous system – all of which I had believed to be true – that mind instead has been there always, and that this universe is life breeding because the pervasive presence of mind had guided to be so".[8] He added, "I speak of mind pervading the universe, of mind as a creative principle perhaps primary to matter".

Fritjof Capra wrote, "By consciousness I mean the state of mind characterized by self-awareness , where one is aware that he is aware".[9] Richard Gregory wrote, "Matter is public but mind is uniquely private".[10]

Wilder Penfield wrote, "To suppose that consciousness or the mind has a location is a failure to understand neurophysiology".[11] Fred Alan Wolf wrote, "We are beginning a new age of awareness, the age of quantum consciousness, the age of conscious atom. By looking within ourselves, we may be able to solve the problems facing us on the final frontier – the frontier of the human spirit".[12]

4. On Truth or Love

J. Krishnamurti, a renowned world philosopher-mystique, observes the conditioned psyche [symbolic-dualistic frame of reference of mind spectrum], it makes a quantum jump to non-dual and non-conceptual frame of reference of Mind Continuum. The Mind eliminates altogether the sense of duality inwardly. There is only the "fact", not the invented opposite. There is the "fact", "what is", the "true", the "non-conflict"; and reject the "non-fact", "what should be", the "false", the "conflict".[13]

Love is not different from truth. Love is that state in which the thought process, as time, has completely ceased. Where love is, there is transformation, or mutation, or revolution.

The fact is defined as: (a) A state of great depth, stillness in which there is no conflict, no contradiction; (b) The fact itself is the operator, not you operating on the fact; (c) If you leave it alone, there is no time. When the fact is left alone completely, it blossoms instantly; (f) Let the fact live; (g) Thought has no place in the exploration of the inner.

Only the truth can liberate the mind from our ideation; to see the truth the mind must realize the fact that as long as it is agitated it can have no understanding. Seeing the false as the false and the true as the true is transforming, because when you see that something is false, the false drops away.[14]

5. The Human Survival

Many eminent scientists have discussed that the Climate Change is linked to human activity. George Wald wrote, "By now the problem is not transcendence but survival".[15] Jacques Monod, a Nobel prize winning biologist, wrote, "What we are opening before us today is an abyss of darkness".[16] P. Krishna, an eminent physicist, wrote, "We are going towards self-annihilation and not too much time is left now".[17] R.K. Pachauri said, "Only a few years are left to avert the worst impacts".[18]

To sum up, the human survival is embedded in manifestation of higher quantum energy potential through: (a) from reason to intuition, and (b) from objective science to subjective science. Subjective science deals with the Wholeness or Oneness of Life. In subjective science both the object of cognizance and the subjective truth are in the inner realm. It is perceptive Sensitivity or Intelligence. Subjective science is independent and synergistic to objective science. Your consciousness is the consciousness of all humanity; and we have to observe and understand that in its Wholeness.

6. On Limitations of Man

Victor F. Weisskopf, a renowned physicist, wrote, "It seems the present-day physicists have lost the gift of foresight inherited from their forerunners .... The man divorced from nature could not be happy and man's power to resist nature is hopelessly puny .... It is a difficult task of turning the age-old tide of human ambition and aggression .... There is a large gap between the superhuman task and the frailties of human endeavors".[19]

Einstein discussed "the nature of God, the universe and man", and said, "We know nothing about it at all. Our knowledge is but the knowledge of schoolchildren – we shall know a little more than we do now. but the real nature of things – that we shall never know, never".[20] He added, "Our task must be to free ourselves from this prison [conditioned mind] by widening our circle of compassion to embrace all living creations and the whole nature in its beauty".

Einstein said, "It is in fact, nothing short of a miracle the modern methods of instruction have not entirety, strangled the holy curiosity of enquiry, for this delicate plant, asides from stimulation, stands mainly in need of freedom, without this it goes to wrack and ruin without fail". Man is a miniaturized cosmos. He is a limited being.

Krishnamurti, is a prophetic vision, wrote, "Man is not the measure of himself .... No man from outside can make you free .... No organisation can lead man to spirituality .... It is the responsibility to begin with ourselves and to bring about the radical transformation within ourselves"[21]

Voltaire wrote, "Everything is governed by immutable laws .... everything is prearranged .... everything is a necessary effect .... We are both toys in the hands of destiny".[22] Herman Weyl wrote, "I have a direct knowledge entirely different from the theoretical knowledge that represents the parallel cerebral process in symbols".[23] Hideki Yukawa wrote that Galileo Galilei taught us that the mere acceptance of stimuli which come from the environment and which result in the accumulation of memory in the brain is insufficient for our purpose. He taught us by his own achievement that experiments purposely designed are of vital importance in uncovering the truth hidden in nature. He asked questions addressed to nature and succeeded in receiving answers directly from nature. In doing so, he also needed rather simple devices".[24]

Erwin Schrodinger wrote, "No personal god can form part of a world model that has only become accessible at the cost of removing everything personal from it .... I do not find God anywhere in space and time ...."[25]

Man has no inbuilt capacity to resolve the challenge of Climate Change or Global Warming. Krishnamurti stated:

(a) Something has been destroyed in India, but a presence waited in the land. Krishnamurti said, *"The benediction of the timeless is waiting, the ground is ready. Can we do something that is of this benediction"*.[26]

(b) One actually touched the source of energy of all things it is not from the mind or brain but the source of itself. *It was a state in which the mind or the brain was no longer in operation only the source was in operation.* [27]

(c) It is the benediction of Nature that heals, but man has to prepare to perceive that Perception.

7. The Nature of God and Man

The God, the Divine Consciousness, the Life Field, an epitome of Truth-Love-Compassion is the creator and sustainer of all that exists. We find ourselves in a universe that breeds life and possesses the very particular properties that make that possible. A presence is waiting in India.

Consciousness expresses itself in different quantum energy potentials – say, conditioned symbolic–dualistic energy and Unconditioned Mind Continuum. The non-dual and non-conceptual frames are intuition, invention and discovery. Nature, out of its benediction, transcends these and the energy of Silence gets awakened; and is also called the Divine Intelligence. The benediction of the timeless is waiting, the ground is ready, and Nature awakens to express its ***Grace, Order and Harmony.***

The Intelligence is beyond Mind and beyond brain, beyond time-space matrices, beyond all tensions and pressures, it is limitless, immeasurable and infinite; it is born out of the creation of Silence. It is the Intelligence that heals because it has the dimension of love and compassion.

States of Mind, Brain and Body System: It is the "subjective-objective observing and understanding" that bestows "peace and serenity" to man. When attention, when the non-evaluatory perception is sustained, when there is non-identification with the emotional roles, when the subjective momentum of identifying and comparison are transcended, when we look at thought in utter freedom, the though structure goes to abeyance. There is nothing like and individual consciousness – there is racial consciousness and a racial momentum of perception.

Man : Our minds and our brains are not at all the same. It is the mind which experiences and it is the brain which records the experience. Matter is public but mind is uniquely private. It seems as if present-day physicist have lost the gift of foresight inherited from their forerunners. All over the world, the mediocricity is increasing. Learning depends on the state of attention. The education is leading us to narrow-minded scientists and philosophers. Your consciousness is the consciousness of all humanity.

All science presupposed the presence of the power of intuition. At present all sciences seem to lose the power of intuition. We must not only emphasize knowledge, but "insight" which has the ability to see a truth directly.

Man is a miniaturised cosmos. His knowledge and experience is limited. He is not the measure of himself. Einstein stated, "Our knowledge is but the knowledge of schoolchildren. The real nature of things – that we shall never know, never".

The climate change is basically due to conditioning the human psyche. This leads to derive crisis of receding glaciers. The derived cause enhances the conditioning of the human psyche at an accelerative pace, and this conditioning is bound to cut down the period of human survival. If humanity is to survive, we have to educate mind-brain-body system, the fundamental crisis, and live in non-violence, peace, freedom, love and compassion.

8. Concluding Comment

(a) If humanity is to survive on this earth, we will have to undertake education of the conditioned mind and Unconditioned Mind with imperative sincerity and truthfulness to *know* the mind, to *shape* the mind, and to *liberate* the mind.

(b) One must be the willing servitor of the Divine Consciousness.

(c) The only solution to end human challenges is higher quantum energy potential of Awareness/Consciousness.

(d) We need "subjective and objective observation and understanding". Unprecedented understanding and perception across the globe is needed to tackle the "dancing challenge" of Climate Change.

(e) Only the truth can liberate the mind from its own ideation. Seeing the false as the false, and the true as true is transformation, because when you see something very clearly as the truth, that truth liberates. When you see that something is false the false thing drops away".

(f) It is our responsibility to begin with ourselves and to bring about the radical transformation within ourselves.

References

1. R.K. Pachauri, *The Tribune of Nov. 17, 2007.*
2. *Ibid.,* The Tribune, Nov. 26, 2007.
3. LA Times, Washington Post, in *The Tribune, February 22, 2008.*
4. The Hindu, *February 28, 2008.*
5. The Tribune, March 7, 2008.
6. J. Krishnamurti, in *The Urgency of Change,* London Victor Gollancz Ltd., 1977, p. 189, 190, 191.

7. P. Krishna, *Education, Science and Spirituality,* USA: California Institute for Higher Educaiton, Research and Dialogue, April 3, 1993, p. 14.
8. George Wald, in *Synthesis of Science and Religion,* Mumbai: The Bhaktivedanta Institute, June 1988, pp. 19-21.
9. Fritjof Capra, *ibid.,* p. 275.
10. Richard Gregory, *ibid.,* p. 285.
11. Wilder Penfield, in Fred Alan Wolf, San Francisco: Harper and Row Publications, 1981, p. 244.
12. Fred Alan Wolf, *ibid.,* 1981, p. 246.
13. J. Krishnamurti, *Within the Mind,* KFI, 1982 pp. 148, 151.
14. *ibid., The First and Last Freedom,* London: Victor Gollancz Ltd., 1977, pp. 278, 286.
15. George Wald, in *The Evolution of Consciousness,* New Delhi: National Publishing House, 1983, p. 14.
16. Lacques Monod, in *New York: Vintage Books 1972,* p. 171; & in *Synthesis of Science and Religion, op. cit.,* p. 127.
17. P. Krishna, *op. cit.,* April 3, 1993, p. 12.
18. R.K. Pachauri, *op. cit.,* The Tribune, Nov. 18, 2007.
19. Victor F. Weisskopf, in Introduction of *Creativity and Intuition,* Hideki Yukawa, 1973, Japan: Kodansha International Ltd., 1973, pp. 10,9,13.
20. Albert Einstein, *The Merry Mind of Albert Einstein,* Readers Digest, July 1973, pp. 133, 134 and Kenneth Brecher, in *Albert Einstein: 14 March 1879 – 18 April 1955,* Nature, Vol. 278, 15 March 1979.
21. J. Krishnamurti, KFI, 1986/3, pp. 1, 3; KFI 1986/1, p. 5; & KFI, 1986/3, p. 16.
22. Voltaire, *Dictionnare Philosophique,* Paris, Garnier, 1954, in Ilya Prigogine and Isabelle Stengers, London: Bantam Books, 1984, p. 257.
23. Herman Weyl, in Ilya Prigogine, *ibid.,* p. 311.
24. Hideki Yukawa, *Creativity and Intuition, op. cit.,* p. 111.
25. Erwin Schrodinger, *What is Life? & Mind and Matter,* Cambridge University Press, 1967, p. 150.

26. Pupul Jayakar, *Krishnamurti: A Biography*, San Francisco: Harper and Row Publication, 1986, p. 380.
27. *Ibid.*, p. 391.

7

The States of Mind

1. The States of Mind

Man has an evolutionary origin at about 2.5 million years ago. In its long and tortuous journey through the corridors of time, humanity stands poised today at the most critical crossroads. It faces a crisis of consciousness. Stark and endemic poverty, gratuitous degradation of the environments leading to Climate Change, or Global Warming and mindless conflicts because of human hubris appear along with unprecedented economic growth and prosperity.

Physicists have came to see that all their theories of natural phenomena, including the 'laws' they describe, are creations of the human mind; properties of our conceptual map of reality, rather than of reality itself. These are parts of the map, not of the territory. The symbolic mind is invariably dualistic and constructs a time-space matrices, which expresses itself as violence, greed, anger, desire, acquisition, and all psychological imbalances.

Violence is an indication of psychological immaturity when the scientific mode of mind sees the cause of psychological imbalances, the mind makes a quantum jump from symbolic-dualistic frame to non-dual and non-conceptual frame of the

Mind Continuum, where the non-dual frame of reference is synergistic or at higher quantum energy potential of Consciousness. The art of ***"Subjective-objective observation and understanding"*** lays foundation of "humanist technology", or "science of life", or "subjective research". To sum up violence is a construct of symbolic-dualistic frame of reference, and non-violence is a percept in the non-dual frame of reference of Mind Continuum.

Hideki Yukawa wrote, "The capability of abstract thought .... always presupposed the presence of the power of *"intuition"* with which both Greek and Chinese geniuses in ancient times were richly endowed. The important point would appear to be the balance or cooperation between intuition and abstraction".[1] It seems as if present-day physicists have lost the gift of foresight inherited from their forerunners".

J. Krishnamurti wrote, "Knowledge is limited. All over the world, mediocrity is increasing"[2] .... Learning depends on the state of attention". Education leads us to narrow-minded thinking especially among scientists: Krishnamurti added, "Invention is based on knowledge".[3] We opine that invention is based on perception and not knowledge.

Hideki Yukawa wrote, "Man discovered new things hidden in nature .... Science can be creative in the sense that it results in the discovery of a truth in nature hitherto unknown to man .... Science is a manifestation of man's effort to open up the unknown world that lies before him, to discover new possibilities".[4]

This article explores "The nature of God, the universe and man". It discusses the states of mind with Consciousness. It discusses the deep meanings of intuition, invention [insight] and discovery. It explores the highest potential of Consciousness which is called the Intelligence, or the Divine Consciousness or the Life Field.

2. What is Mind?

Eugene Wigner, Nobel Prize in physics, 1963, wrote, "Self-identity demands consciousness. You would not have identity without consciousness. Yes, consciousness creates identity. I know that I am I. And it is my consciousness which knows, which tells me that".[5] Fritjof Capra wrote, "By consciousness, I mean the state of mind characterized by self-awareness, where one is aware that he is aware".[6] Richard Gregory wrote, "Matter is public but mind is uniquely private.[7]

Wilder Penfield, a Canadian brain surgeon, wrote, "Our minds and our brains are not at all the same .... It is the mind which experiences and it is the brain which records the experience".[8] George Wald wrote, "Mind has been there always and this universe is life-breeding because the pervasive presence of mind has guided it to be so".[9] Mind is pure consciousness and it has no specific location. It expresses itself at the quantum energy potential of Consciousness. We have no way of identifying either the presence or absence of consciousness, let alone its contents. Language can model the states of the world, but not states of mind. Mind is not only locatable, it has no location. It is not a thing is space and time, not measurable.

3. A Meditative Way of Life

Einstein, in a prophetic vision, stated that we need "objective observing and understanding". Observing is only possible when the subject is attention and that non-evaluatory perception is sustained.

When we are looking at life as consisting of movement and non-movement, motion and non-motion, action and non-action, these two together make the wholeness of life. We see that the movement of thought-structure is independent of my volition, it is independent of the movement of the ego. The movement of thought structure is not a voluntary movement at all; it has nothing to do with his will, nothing to do with his desires, it has its own independent movement. A movement of thought is equal to tension in your whole nervous system. And a movement of emotion and sentiments is equal to chemical turbulence in the body. There is sheer motionlessness. No thought moves, no feeling no sentiment, no emotion is moving.

Meditation is a way of living where the relationships are movements of relaxation and not tension and pressures.[10] Silence is a tension-free and pressure-free state of your being.[10] This exploration tells us that there is racial consciousness and a racial momentum of perception, experience and thought, and there is nothing like an individual ego, the self the me, there is a sense of great freedom within.[11] This leads us "through silence to meditation". It is an enquiry on behalf of the whole human race.

4. The States of Mind-Brain-Body System

Intuition: You give a small child a toy. The child absorbs it, takes him over. It you see a great marvellous snow-peaked mountain, the depth of the mountain, the grandeur, the magnificence, the glory of it. The mountain has taken you over. That means that a great thing has absorbed you. If an 'old thought' momentarily absorbs you and is replaced by a 'new thought', which is at a higher potential of consciousness, it is called "intuition".

It is a state where everything is washed out without the mountain or the gorge. There is the absence of the self, the "me", the "I-consciousness". Meditation implies that the Unknown – a fathomless and boundless, ocean – absorbs the non-dual Mind [the subject as attentivity and the object as the fact]; or in other words, the subject and the object is in the non-dual state of mind; both coalesce and both vanish, and what is left behind in pure Consciousness – a state of great depth, stillness in which there is no conflict, no contradiction.

A symbolic-dualistic frame of reference of mind cannot do any meditation as the mind is repetitive and mechanical. It is the non-dual and non-conceptual frame that is responsible for meditation.

Invention: It is based on non-dual and non-conceptual frame of reference of Mind Continuum. It is not-based on mutilated perception. The distorted perception means the non-fact, the false, the past, the what should be?; The pure perception implies the fact, the true, the present, the what is! Some philosophers call the invention as insight. Insight implies not having the continuity of memory which is the past; but the enduring of it and seeing something new.[10] Insight is to be free of conditioning of the brain and the total freedom gives you a complete holistic insight.

Discovery: Hideki Yukawa discussed the essence of discovery.[11] Krishnamurti discussed that creation is not something continuous, it is not something in the end.[12]

Sensitivity is conferred upon the person by the perception of truth. Mahatma Gandhi wrote, "Truth is my God, Truth is my defence, Truth is my strategy". Gandhi had put something in the orbit of global human consciousness and even today what he had set into motion, in still alive.[13]

5. Intelligence

The Urge for Freedom: Understanding the nature of bondage is the beginning of freedom. Understanding the content of bondage gets converted into a factual urge for freedom. A non-subjective perception is the foundation of the religious life, which implies non-dual, timeless, direct perception. We ourselves are part of the cosmic dance of energy.

The Benediction of Life: Krishnamurti felt the sacredness of Indian land. A presence waited is the land. He said, "The benediction of the timeless is waiting, the ground is ready. Can we create something that is of this benediction".[14] He conveyed a way of perception, of challenge and response from a state that lay beyond the mind, beyond brain, memory, all responses of consciousness. When this state is attained, the non-personal, the non-individuated energy gets activised; and this is often called the energy of Intelligence, the Divine Consciousness, the Life Field. The Intelligence is free from all conflict, all chaos and psychological imbalances.

Man is a miniaturized cosmos, he carries the perception of all humanity, all the energies existing in the cosmos are within you. The movement of the Mind is the movement of the total human race.

That tension-free, pressure-free state, that division-free state, that duality-free state, activises the energies of the Wholeness. Intelligence is the energy of Wholeness. Intelligence is the energy of that complex inter-relatedness of life. It is the benediction of life to all existences on this planet and beyond.

The Intelligence [Active Reality] is a powerful, highly directional, monochromatic [same frequency] and coherent [the wave is in one phase] beam of energy; or a state of highest frequency [higher frequency means higher energy], or higher quantum energy potential. The intense energy of Intelligence expresses itself as direct expression of the fact, the Reality. The Intelligence acts upon the brain matter and transmutes it, so the brain matter transforms and attains a higher order through Intelligence. The understanding permits the powerful subtlest energies to penetrate the known and transforms the known. The subtlest energy [higher potential] knocks the thought out of commission, not by thinking, not by reasoning, but rather a direct transformation due to its contact with the subtlest energy. The Insight is the supreme Intelligence.[15]

6. A Critique on "Climate Change"

The UN: IPCC must engage itself in a fundamental challenge *[mystic intuition]* and a derived challenge *[scientific analysis]*. It is a crises in our Consciousness.

Recently it has shown that the area of glaciers has been reduced from 3,391 to 2,721 sq. km. between 1962-2004. They reported that the overall reduction of 21 percent in the glacial surface area. The mean surface area of glacial extent is from 1 sq. km. to 0.32 sq. km. during 1962-2004.[16] More recently, it has been reported in the Parliament that 35 glaciers shows that 43 percent of glaciers in the Himalayan region are receding in the range of 10 to 20 m/year.[17] In a dialogue at Chandigarh on March 06, 2008, we discussed "The Immanent Disaster" or "A Vision on Climate Change".[18]

The fundamental crisis is the psychodynamics of our conditioned psyche; or the symbolic-dualistic frame that constructs the conditioned energy; or the energy of violence; or the psychological imbalances. The conditioned energy [violence-Unmanifest-violence] travels in the external matrices. The derived crisis is when a glacier residing on some rock deep down – say a glacier resting on some basic foundation – starts releasing the stored energy in melting of the glacier; or in other words, receding of the glacier. As the cause of recession of glaciers is not known, so the measures of reducing the recession are totally unknown.

The "scientific analysis" or the "derived challenge" suggests that it is a happening at the deepest level of glacier because the receding started. The conditioning of human psyche is increasing at a rapid rate and this conditioning is bound to cut down our survival period.

The answer to this paradox is to end the human conditionings through higher quantum energy potential of Awareness / Consciousness. It is the holistic education of the *"subjective-objective observing and understanding"*, or holistic education of the Wholeness of Life.

It is imperative for every human being, for all political parties, media and governments to cut down all causes leading to rising temperatures, melting glaciers and extreme weather events across the world. Human-induced Climate Change adds new stresses – as nervous tensions and chemical pressures. We want new relations between man and man. We need a dynamic interplay of objective science and subjective science [spirituality].

The Intelligence [Divine Consciousness] heals, because it has the dimensions of creativity and spontaneity, non-violence, beauty and harmony, love and compassion. Man has to choose between healing and annihilation, between the Buddha [Wisdom] and the bomb [annihilation]. Obviously the decision in to work honestly, sincerely and truthfully mutation of one's man psyche.

7. Concluding Comment

The UN: IPCC had unequivocally affirmed the warming of the climate system and linked it directly to human activity. It must engage itself in a fundamental challenge *[mystical intuition]* and derived challenge *[scientific analysis]*. It is a crisis in our consciousness. We need a higher hierarchical level of Awareness / Consciousness to resolve the fundamental Challenges of consciousness.

The receding of glaciers is basically due to "derived challenge", which increases the conditionings at a rapid rate. This conditioning is bound to cut down our survival period. It is a serious challenge.

It seems as if the present-day physicists have lost the gift to foresight inherited from their forerunners; science is losing the power of intuition; all over the world, the mediocrity is on the increase; learning depends on the state of attention; and all these trends combine to destroy the unity of man's various aspects and leads eventually to the loss of his humanity. It is our responsibility to begin with ourselves and to bring about the radical transformation within ourselves.

Reference

1. Hideki Yukawa, *Creativity and Intuition,* Tokyo: Kodansha International Ltd., 1973, pp. 101-102, 105.
2. J. Krishnamurti, *KFI* 1986/2, pp. 24, 17.
3. *Ibid.,* KFI 1986/1, P. 17.
4. Hideki Yukawa, *op. cit.,* pp. 112, 113, 205.
5. Eugene Wigner, in *Synthesis of Science and Religion,* Mumbai: The Bhaktivedanta Institute, June 1988, p. 262.
6. Fritjof Capra, *ibid.,* p. 275.
7. Richard Gregory, *ibid.,* p. 285.
8. Wilder Penfield, in Deepak Chopra, *How to Know God,* USA: Rider, 2000, pp. 215, 217.
9. George Wald, in *Synthesis of Science and Religion, op. cit.,* pp. 19-20, 18.
10. Vimala Thakar in *Silence,* 1989; & Through Silence to Meditation, Ahmedabad: Vimal Prakashan Trust, 1998.
11. *Ibid,* 1998, p. 27.
12. J. Krishnamurti, *KFI* 1985/3, p. 30.
13. Sampooran Singh, *Education for a Global Society,* Dagshai Cantt., Atam Science, 7.10.2001.
14. J. Krishnamurti, in Pupul Jayakar, *Krishnamurti: A Biography,* 1986, p. 380.
15. Sampooran Singh et. al., *God in Science,* Vishvas Publications Pvt. Ltd., Chandigarh, July 1999, p. 2.
16. R. Ramachandran, in *Stronger Evidence of Global Warming,* The Hindu, February 28, 2008.
17. Visha Sharma, in *Himalayan Glaciers Receding;* Meena; The Tribune, March 07, 2008.
18. Sampooran Singh, *The Immanent Disaster,* or *A Vision on Climate Change,* A Dialogue on March 06, 2008.

8

The Ending of the Climate Change

1. Introduction

The UN: IPCC Chairman R.K. Pachauri, in November 2007, had "Unequivocally affirmed the warming of our climate system and linked it directly to human activity". He stressed that only a few years are left to avert the worst. The objective science reveals the magnitude, geographic extent and precise locations of ocean warming.

This article discusses the cause of "climate change" and also explores the remedial measures to avert the crisis. The UN: IPCC must engage itself in a fundamental challenge *[mystical intuition]* and derived challenge *[scientific analysis].* It is a crisis in our consciousness.

2. The Cause of the Challenges

The Prime Minister Dr. Manmohan Singh inaugurated ***"A National Plan of Action for Climate Change"*** on February 7, 2008. The basic plan deals with socio-economic-political aspects, which deals with material prosperity. It deal with poverty reduction of the poor people everywhere. The plan is silent on higher potentials of Awareness/Consciousness.

Objective science deals with discovery of a truth in nature hitherto unknown to man. Science is a manifestation of man's effort to open the unknown world that lies before him, to discover new possibilities. There are two important dimensions of science – one is the method of *induction* which deals with the inner world, and the other is the method of *deduction* which deals with the outer space-time matrices. Depending on the state of mind-brain-body system, we may classify the challenge in three states – (a) the basic challenge, which deals with the conditioned mind-brain-body system; (b) the fundamental challenge, which deals with the Unconditioned Energy; and (c) the derived challenge, which deals with energy in the outer space-time matrices. Based on this definition, we conclude that the climate change is due to the fundamental challenge in the inner field of man, and the derived challenge at the terminal end of melting glaciers.

Reforms: The national plan is a set of reforms. Hideki Yukawa wrote, "The need for a reform is not without, but within".[1] A reform needs a mutation. Krishnamurti wrote, "Reforms needs further reform, and this is an endless process".[2] Reforms enhance conditioning the human psyche. Krishnamurti added, "The only revolution is the revolution between man and man, between human beings. That is our only concern".[3]

Loss of Foresight: Hideki Yukawa wrote, "It seems as if the present-day physicists have lost the gift of foresight inherited from their forerunners .... This seems to be related to the theme of intuition and abstraction".[4] Maurice H. Wilkins wrote, "Science makes progress largely by a combination of two aspects: open-mindedness and the following of abstract concepts. Abstraction involves excluding and narrowing down, which is very opposite of open-mindedness. Science always puts the most emphasis on this abstracting and narrowing down, on analytical procedures and on the role of the intellect as distinct from intuition and feelings".[5] We unequivocally affirm that objective science cannot resolve any psychological imbalance and we need mystical intuition and scientific analysis.

Victor F. Weisskopf wrote, "Abstraction cannot work by itself, but has to be accompanied by intuition".[6] Hideki Yukawa wrote, "It always presupposed the power of intuition". Physicists lay a stress on abstraction which results in loss of intuition, or creative thinking. Krishnamurti wrote, "All over the world, mediocrity is increasing."[7] Loss of gift of foresight, loss of intuition, mediocre living add to the danger of conditionings and eventually loss of humanity".

Krishnamurti wrote, "Man is not the measure of himself .... No man from outside can make you free[8] .... No organization can lead man to spirituality".[9] Thought is sterile to bring any transformation in the conditioning of the human psyche. All the things that man has put together have to be totally abandoned.[10] This means that thought – because it is repetitive and mechanical – is sterile in bringing any transformation in the human psyche. We have to educate our mind-brain-body system so that it is rejuvenated to perceive non-violence and be free of all psychological imbalances.

Observation: It is in a non-dual and non-conceptual frame of reference of Mind Continuum. There is no duality. Observation without an observer is attentivity. It is subject-free and object-free attentivity. In attentivity there is understanding. Understanding releases the state of Awareness / Consciousness.

Seeing, hearing, smelling, breathing are direct perceptions, but thinking in the symbolic-dualistic frame is indirect experience. The direct movement is in the realm of non-duality.

Education: Maurice H. Wilkins wrote, "I agree with Einstein, that the sort of scientific education we have now has produced a narrow-minded way of thinking among scientists, so that they give no proper attention to the moral and psychological dimensions .... Science as it is today does not tell us very much about the real nature of human beings".[11] The objective science leads to enhancing the *"derived causes"*. The present experimental data of receding glaciers unequivocally affirm that all "derived causes" combine to destroy the unity of man's various aspects – including all hidden and unknown variables – and all lead eventually to the loss of his humanity. Such a split will deprive man of peace, freedom, love and compassion.

We define education as liquidating the imbalances and stimulating spontaneous equipoise and equibalance.

Insight: P. Krishna wrote, "A mutation is possible in the brain through insight, not through knowledge, not through accumulation. That insight is the ability to see a truth directly, direct perception of the truth, and education must foster that. So we must not only emphasize knowledge but emphasize insights – the possibility of insights".[12]

When attention is active when that non-evaluatory perception is sustained, when there is non-identification with the functional roles, when the subjective role of identifying, comparison are transcended, when we look at thought in utter freedom, then thought-structuring goes to abeyance. The absence of thought-structure results in regeneration of Silence.

3. Perception of Creativity

Creativity implies discovering something that nobody has hitherto known about or inventing something new. Invention and discovery always have something of the nature of unpredictable. Our understanding or perception of natural phenomena, makes a great leap forward.[13]

Wilder Penfield, a Canadian brain surgeon, wrote, "Our minds and our brains are not at all the same .... It is the mind which experiences and it is the brain which records the experience".[14] The Mind – which is energy itself – receives the energy from the Unmanifest [Life Field], and passes it on to brain and body. The Mind-Brain-Body works as one system at a given Awareness / Consciousness; and constructs *intuition, invention* and *discovery,* which are at different quantum energy potentials. A brief discussion on intuition, invention and discovery have been discussed earlier.[15]

4. Where there is a Cause, there is an end to that Cause

The movement of thought structure is independent of my volition, it is independent of the ego. Every inner movement is the movement of the inheritance, the past, every inner movement is conditioned by the past and there is no freedom in the dimension of that movement. Thought-structure knows no freedom. Cerebral movement is irrelevant to learning.[16]

You have to be attentive at looking at life now as consisting of movement and non-movement, motion and non-motion, action and non-action, the identification with functional roles is transcended. There is shear motionlessness. No thought moves, no feeling, no sentiment, no emotion is moving. Meditation is a way of living where relationships are movements of relaxation and not tensions and pressures.[16] It tells you that: (a) Thinking is the movement of all life, it is not individual thinking; (b) You move away that it is not my thinking, it is global thinking; (c) Your brain is not yours, it is evolved through thousand and thousand years; and (d) There is nothing like an individual mind, there is nothing like an individual consciousness. When all this is seen that there is racial consciousness and a racial movement of perception, there is a sense of great freedom within.[17] Freedom leads to ending the cause.

5. Perception of the Fact

Krishnamurti wrote, "We have lost in India, a sense of sensitivity to the world outside of us".[18] It implies that we are also not aware of our inner life. Perception of the fact implies to "observe and understand" the wholeness of life; one has to be aware of the *fact,* the *true,* the *present,* the *what is!* and jettison the *non-fact,* the *false,* the *past,* the *what should be!*

Lord Buddha said, "Self is, Truth is. Where self is, Truth is not. Where Truth is, self is not. And once 'self' dissolves, we have true freedom, freedom from time, freedom from all our sufferings ...."[19] It is a challenge for our education. What the Buddha, proclaimed about the imperative need to *know* the mind, to *shape* the mind, and to *liberate* the mind has been never more true than in the age of the Atom".[20]

Krishnamurti wrote, "To be without the Self, in that state, there is great beauty. Where the self is there is no beauty .... You have to watch [observe] a state of great depth, stillness in which there is no conflict, no contradiction .... The fact itself is the operator, not you operating the fact .... When the fact is left alone completely it blossoms instantly .... Let the fact live".[21] He added, "Thought has no place in the investigation of the inner .... There is no division between the thinker and the thought, the experiencer is the experience, the observer is the observed – this will eliminate duality,[22] and will bestow the perception of the fact.

6. Transformation

Mutation, transformation, revolution, are synonymous words. It is not a radical change in the quality of the mutant. It is abeyance of the conditioned mind. It is not a peripheral change. It is change in the whole mutant.

Transformation is not in the future. It can only be now, from moment to moment. Seeing the false as the false, and the true as the true is transformation, because when you see something very clearly the truth, that truth liberates. Truth is not cumulative. It is from moment to moment. That which is cumulative, it is memory.[23]

Only the truth can liberate the mind from its own ideation; to see the truth, the mind must realize the fact that as long as it is agitated it can have no understanding. Stillness of mind comes only when there is no process of isolation through accumulation but complete understanding of the whole process of relationship. In stillness there is no projection of thought.[24]

Love is not different from truth. Love is that state in which the thought process has completely ceased. Where love is, there is transformation.

7. Ending of Thought

Thought, at all levels, is obviously the reaction of memory. It emerges from a symbolic-dualistic frame of reference of mind spectrum. It constructs space-time matrices and is recorded in the brain as conditionings. Every thought creates a tension and every emotion upsets the equilibrium of the chemical system. Thought in repetitive and mechanical, so it has no originality.

Albert Einstein, in a prophetic vision, wrote that we need, *"Objective observing and understanding"* to achieve *"peace and serenity"*. Scientific observing implies to observe non-subjectively, to observe without any evaluatory comparisons, or judgement creeping in. I look at the thought-structure – not to judge, neither to accept or reject, nor to act upon them, not to change them. I look at the movement of thought in utter freedom. This leads to non-dual and non-conceptual frame of reference of Mind Continuum. In other words, the Mind becomes receptive or sensitive or attentive. In an attentive Mind, thought goes to abeyance.

The mind is structured in different levels of Awareness / Consciousness, and each level is synergistic to the previous level – the levels are intuition, invention and discovery, and each level transcends thought. If this observation is continued for a long time, the Divine Intelligence or Life Field gets activised which resolves all problems / challenges, and man reveals in pure Consciousness.

8. A New Way of Living

By allowing a psychic mutation to occur in our being, we would be reducing one area of violence from the globe. This will also lead to reducing other psychological imbalances. By offering our life for that psychic mutation to take place, we might be creating a living cell of a new life style, new way of living. We might be setting into the orbit of human consciousness a spark of love and compassion.[25] We embark upon exploration on behalf of the whole human race. You turn your face towards truth, towards the homogeneous wholeness of life, towards the mystery of life. The intelligence provides the help when it is really needed.

It is the Universal Consciousness, the Life Field that can end the climate change, and lead mankind to survival, love and compassion.

9. Concluding Comment

The UN: IPCC has unequivocally affirmed the climate change. We opine that this is due to fundamental challenge in the inner world of man and derived challenge at the terminal end of melting glaciers. The UN: IPCC must engage itself in the fundamental *[mystical intuition]* challenge and derived *[scientific analysis]* challenge.

The derived challenge basically is due to the reforms, loss of foresight, loss of intuition, mediocre living, scientific technology, and education leading to narrow-mindedness. This leads eventually to loss of humanity and integrity. These result in accumulation of conditioning, which lead to decrease of our survival period. It seems that the melting of glaciers is increasing everywhere and everyday. An example is that

Antarctica's Wilkins Ice Shelf [South Pole] is just tumbling and crumbling in the ocean.[26]

We will have to regenerate insight and perception of creativity. We need education of the mind. We have to be aware of the *fact* the *true* the *present*, the *what is!* If one "observes and understands" a cause, there is an end to that cause. There is nothing like an individual "I-consciousness". Thought is product of collective human effort through centuries.

We have to educate our mind that there is racial consciousness and a racial momentum of perception. We need a higher hierarchical level of Awareness / Consciousness to resolve the fundamental challenge; and the derived challenge on climate change will vanish on its own. It is a collective exploration on behalf of the whole human race.

References

1. Hideki Yukawa, *Creativity and Intuition,* Tokyo: Kodansha International Ltd., 1973, p. 124.
2. J. Krishnamurti, *The Urgency of Change,* London: Victor Gollancz Ltd., 1977, p. 189.
3. *Ibid.,* p. 190.
4. Hideki Yukawa, *op. cit.,* pp. 105, 106.
5. Maurice H. Wilkins, *Synthesis of Science and Religion,* Mumbai: The Bhaktivedanta Institute, June 1988, p. 30.
6. Victor F. Weisskopf, in Introduction, Hideki Yukawa, *op. cit.,* pp. 10, 11; & Hideki Yukawa, *op. cit.,* p. 101.
7. J. Krishnamurti, KFI, 1986/2, p. 24.
8. *Ibid.,* KFI, 1986/3, pp. 1, 3.
9. *Ibid.,* KFI, 1986/1, p. 5.
10. *Ibid.,* KFI, 1986/1, p. 14.
11. Maurice H. Wilkins, *op. cit.,* p. 35.
12. P. Krishma, *Education: Science: Spirituality,* California: Institute of Higher Education, Research and Dialogue, April 3, 1993, p. 14.

13. Hideki Yukawa, *op. cit.,* p. 123.
14. Wilder Penfield, in Deepak Chopra, *How to Know God,* London: Rider, 2000, pp. 215, 217.
15. Sampooran Singh, in *The Nature of God, The Universe, and Man & A Vision on Climate Change,* A Report submitted to R.K. Pachauri, March 18, 2008, pp. 12-13.
16. Vimala Thakar, *Through Silence to Meditation,* Ahmedabad, 1998.
17. J. Krishnamurti, *Mind Without Measure,* KFI, 1982-83, pp. 102-109.
18. *Ibid., Within The Mind,* 1982, KFI, p. 147.
19. Lord Buddha, in B.M. Khanna, *The Odyssey of Science, Culture and Consciousness,* New Delhi: Abhinav Publications, 1990, p. 165.
20. D.S. Kothari, in Atom and Self, *The Evolution of Consciousness,* New Delhi: National Publishing House, 1983, 239.
21. J. Krishnamurti, *Within The Mind,* KFI, 1982, pp. 147-149.
22. *Ibid.,* 151-152.
23. J. Krishnamurti, *The First and Last Freedom,* London: Victor Gollancz Ltd., 1977, p. 286.
24. *Ibid.,* p. 278.
25. Vimala Thakar, *op. cit.,* 1998, p. 49.
26. Ted Scambos, Washington, *Ice Shelf Collapses Amid Warming,* The Hindu, March 27, 2008.

9

The Ending of the Human Challenges

1. Introduction

We are living in one of the most critical periods of human history when all the intellectual and emotional currents of all the people of the world over converge upon one and the same point. Man succeeded in discovering reason and developed through many centuries. Reason developed to time and he developed the idea of sequence and continuity.

We are living in the age of science. Objective science has certainly contributed towards the progress and well-being of mankind in many fields – agriculture, medicine, super-sonic travel, and so on. Ironically, however, increasing visible side-effects of industry and technology confront us with grave dangers as environmental hazards, shortage of natural resources and nuclear war. These indicate an accelerated erosion of human values and of ethical standards leading inevitably to the nullification of the achievements of modern science. These lead to conditionings – symbolic-dualistic patterns – which are repetitive and mechanical and have no originality. The economic, political, social and cultural life is full of violence, conflict and contradiction.

A generation is coming up determined to explore the dimension of consciousness, a form of energy that is entirely different from the brain and its product thought, knowledge and experience. This explores the dimensions of consciousness with a vision to end all human challenges – as violence, conflict, fear and climate change. It also explores the movement of life going on incessantly around me and within me.

2. What is Life?

Life is a homogeneous wholeness. Life is a multi-dimensional phenomenon and human beings who are expressions of that phenomenon are themselves multi-dimensional creatures. Life is self-generated, self-controlled, self-regulated phenomenon. In man there is a dimension of immanence and transcendence of Consciousness. The reality is the present, that which is here and now. Every man is a *"toy"* in the hands of destiny.

The universe is governed by immutable laws of Divine Consciousness, everything is prearranged and everything is a necessary effect.

3. Education

In the words of Ramana Maharishi, "Illiteracy is ignorance and education is learned ignorance. Both are ignorant of the true Aim".[1] The "true Aim" is the non-dual mode of knowing while being literate or illiterate concerns the dualistic and symbolic mode of knowing. Ignorance, in other words, is ignorance of the non-dual and non-conceptual mode of knowing, which would instantly reveal the universe to the mind only.

Observation: The symbolic-dualistic frame of reference of mind spectrum engages itself in projection of thought, the past, the non-fact, the what should be! Thought is repetitive, mechanical and has no originality. It is incapable of any observation.

When the mind is attentive, the mind makes a quantum jump to non-dual and non-conceptual frame of reference of Mind

Continuum. When attention, when that non-evaluatory perception is sustained, when there is non-identification with the fundamental roles, when the subjective momentum of identifying, comparison are transcended, when we look at thought in utter freedom, the thought-structure goes to abeyance. I observe the thought-structure, neither to accept nor to reject, not to act upon them, not to change them. I look at the movement of thought in utter freedom.

Understanding: Mental movement cannot lead me to non-duality. It cannot lead me towards totality. A mental movement sees a part of life. A cerebral movement is irrelevant as far as the totality of life, understanding of life; is concerned.

Understanding happens in a non-dual and non-conceptual frame of reference of Mind Continuum. Understanding is the perfume of communion, a direct insight into life. Knowledge you can acquire, understanding you cannot acquire. It is happening in which the total being is involved. To observe and to understand are free of the past. Both deal with the present, the fact, the true, the what is!

Subjective Dimension of Man: Roger Sperry, Nobel prize in physiology and medicine, 1981, wrote, "It gives full recognition to the primacy of the inner conscious awareness as a causal reality".[2] Both science and spirituality follow the law of non-authoritarian perception – an approach not conditioned by the authority of the past.

Mind as Immortal and Eternal: Larry Dossey wrote, "Non-local mind was infinite in space-time. Consciousness is non-local and infinite, therefore Immortal and Eternal. Consciousness is both trans-spatial and trans-temporal, it is not in space and time.[3] Science must explore the human subjective experience.

Insight: P. Krishna wrote, "Such a mutation is possible in the brain through insight, not though knowledge, not through accumulation. That insight is the ability to see a truth directly, a direct perception of the truth, and education must foster that. So we must not only emphasize knowledge but emphasize insights – the possibility of insight".[4]

4. A New "Instrument of Thought"

(a) David Bohm, an eminent physicist, wrote, that we need a new "instrument of thought", which would radically alter the consciousness of the observer, reorientating it towards a perception of the "unbroken wholeness" of which everything is a form". He added, "We must turn physics around. Instead of starting from parts and showing how they work together, we start with the whole".[5]

It appears that physics must start with the whole [subjective dimension of Life], and then explain the objective parts in the light of the Wholeness. The inner dimension is the whole, while the outer, dimensions are the parts. It is the Wholeness that heals, where the healing implies ending of all psychological imbalances, and there is perfect freedom and peace.

(b) ***Consciousness:*** Erwin Schrodinger wrote, "Consciousness is never experienced in the plural, only in singular .... Consciousness is a singular of which the plural is unknown".[6] Universal consciousness is the Wholeness or Oneness. And everything that exists is an expression of consciousness at lower quantum energy potentials.

5. The Ending of Violence or Psychological Imbalances

Violence is an indication of psychological immaturity. ***Violence is a construct by symbolic-dualistic frame of reference of mind spectrum, and nonviolence is the percept in the non-dual and non-conceptual frame of reference of Mind Continuum.***

Einstein, in a prophetic vision, said that we need ***"objective observing and understanding"***, where both observing and understanding are embedded in the non-dual and non-conceptual frames of reference of Mind Continuum. If you sit to relax and look at the movement of thought in utter freedom, then the energy of attentivity becomes activised. You cannot impose nonviolence on a consciousness which is violent because that violence has deep-rooted causes in the psyche. Useless these causes are transcended, there is no ending of violence.

The mind and the brain are independent; we need restructuring of science to encompass the subjective and the objective dimensions of life. Science is a human activity, it must take the ethical values in its ambient. We have to jettison human reason and intellect, its information [knowledge] and experience. And imbibe intuition.

The basic challenge is inner aggression, hatred and anger in the consciousness. Now that disorder has a cause, and if that cause is seen, understood and eliminated, it is possible to end the disorder. You cannot impose the order on top of disorder. The ending of disorder in our consciousness is virtue. There is no such thing as cultivation of a virtue. Virtue cannot be practiced. We must – through observing, through understanding – end the disorder in our consciousness; then there is automatically virtue.

6. A Vision of Thought Structure

The movement of thought-structure is not a voluntary movement at all; it has nothing to do with his desire, it has its own independent movement. It is independent of my volition, it is independent of the movement of ego. My thoughts, my feelings, my memory have a movement and a momentum independent of me, the ego. I look at thought – not to judge, neither to accept nor to reject, not to act upon them, not to change them; so I look at the movement of thoughts in utter freedom, in utter attentivity, in utter awareness. In such observing, thought goes to abeyance and there is understanding. So when it is seen that there is a racial consciousness and racial momentum of knowledge, experience and thought and there is nothing like an individual ego, the self, the "me", there is a sense of great freedom.

7. Mind-Brain-Body System

Hideki Yukawa, a Japanese Nobel Laureate, wrote, "it seems as if present-day physics have lost the gift of foresight from their forerunners .... All science presupposed the power of intuition .... At present all science seems to lose the power of intuition".[7] Krishnamurti wrote, "All over the world the mediocrity is increasing".[8] Learning depends on the state of attention – a higher quantum energy potential of consciousness. The education is leading us to narrow-minded scientists and philosophers. This basically leads to the conditioning of the human psyche at an accelerative pace; and the conditioning is bound to cut down our survival period. This is a very serious challenge as the "derived challenge" [at the terminal end of a melting glacier] is enhancing every moment and every day. We need a higher hierachical level of Awareness / Consciousness to resolve our fundamental challenge; and the derived challenge will vanish on its own. It is a collective exploration on behalf the whole human race.

8. Man

Man is a miniaturized cosmos, an epitome of Truth-Love-Compassion. He is the creator and sustainer of all that exists. Our minds and brains are not at all the same. It is the mind which experiences and it is the brain which records the experience.[9]

Albert Einstein stated, "We know nothing about it [God] at all. Our knowledge is but the knowledge of schoolchildren .... We shall know a little more than we do now, but the real nature of things – that we shall never know, never".[10]

Our task must be to free ourselves from the prison by widening our circle of compassion to embrace all living creatures and the whole nature in its beauty".[11] Einstein added, "It is in fact, nothing short of a miracle that the modern methods of instruction have not yet entirely strangled the holy curiosity of inquiry, for this delicate plant, aside, from stimulation, stands mainly in need of freedom; without this it goes to wrack and ruin without fail".[12] Voltaire wrote, ".... everything is governed by immutable laws .... everything is prearranged .... everything is a necessary effect .... Both are toys in the hands of destiny".[13]

Paul Davies, an eminent physicist wrote, "The laws which enable the universe to come into being spontaneously seem themselves to be the product of exceedingly ingenious design. If physics is the product of design, the universe must have a purpose, and the evidence of modern physics suggests strongly to me that the purpose includes us".[14]

The universe and all men are put up as "toys" by God, which are governed by immutable laws. In the universe, a particular event is a "happening", where everything is prearranged and everything is a necessary effect. All men are miniaturized cosmos, and everyman has an inbuilt capacity to educate his mind-brain-body system and to grow in a vertical dimension. It is up to man either to grow in a vertical dimension, or to degenerate as a plant. He is a toy in the hands of destiny.

9. The Divine Energy of Consciousness

God, Life Field, Divine Energy are Immortal and Eternal. Science must explore the human subjective experience, the inner conscious awareness. So man has to explore his own mind-brain-body system. It is the higher hierachical potential of consciousness that can resolve all human challenges. If the fundamental challenges are resolved, all derived crisis, psychological mind, and climate change will vanish by itself.

Science now recognizes the truth of the immanence and the transcendent potential of each human consciousness, and that of the divine universal consciousness to be the truth of the universe. If man perceives this truth, the boundaries of thought-time and the limitations of the mind-brain-body are divorced and one revels in the realm of intelligence, Wisdom, Love and Compassion. The perception of *Truth* ends all challenges and one revels in freedom, order, non-violence and peace.[14] This is the key to ending of all psychological imbalances, or conditionings, or inheritance.

10. Birth and Death

Between the two points of birth and death all the things with forms exist and operate. Life seems to move on through birth, growth, decay and death. Birth is the emergence of a conscious point and death is the extinguishing of the conscious point. There are immutable laws of nature.

Man has to enhance the quantum energy potential of consciousness through "objective-subjective observing and understanding". Throughout the life span, man has to educate his mind-brain-body system.

References

1. Ramana Maharshi, in Ken Wilber, *The Spectrum of Consciousness,* USA: The Theosophical Publication House, 1982 [Quest Book], p. 110.
2. Roger Sperry, in Wills W. Harman, *Synthesis of Science and Religion,* Mimbai: The Bhaktivedanta Institute, June 1988, p. 199.
3. Larry Dossey, in *Vision and Values,* Lalit K. Kothari and Ramesh K. Arora, New Delhi: Paragon International Publishers, 2006, pp. 49-58.
4. P. Krishna, *Education, Science and Spirituality,* USA: The University of California at Santa Barbara, April 3, 1993, p. 14.
5. Gary Zukav, *The Dancing Wu Li Masters,* Fontana/Collins, 1979, pp. 326, 323.
6. Erwin Schrodinger, *What is Life? & Mind and Matter,* Cambridge University Press, 1967, pp. 94, 95.

7. Hideki Yukawa, *Creativity and Intuition,* Tokyo: Kodansha International Ltd., 1973, pp. 105, 106, 101.
8. J. Krishnamurti, KFI, 1986/2, p. 24.
9. Wilder Penfield, in Deapak Chopra, *How to Know God,* London: Rider, 2000, pp. 215, 217.
10. Albert Einstein, *Reader's Digest,* July 1973, pp. 133, 134.
11. *ibid.,* in Kenneth, Brecher, *14 March, 1879 – 18 April 1955,* Nature, Vol 278, 15 March 1979, p. 216.
12. Voltaire, Dictionnaire Philosophique, in Ilya Prigogine, *et al., Order out of Chaos,* London: Bantam Books, April 1984, p . 257.
13. Paul Davies, *Superforce,* New York: Simon and Schuster, 1984, p. 243.
14. Dr. K. Kaur, P. Singh and S. Singh in *Science with Creativity,* Vedanta, UK: Ramakrishna Vedanta Centre, November – December 2000, No. 296, p. 267.

Section-II
Creativity and Innovation

10

The March of Objective Science

The temper of science is undergoing a continuous transformation. It is now becoming more flexible, dynamic, ecological and holistic. We find ourselves in a universe that breeds life and possesses the very particular properties that make that possible.

Mankind has accomplished magnificent feats in science and technology and brought about dramatic transformation in diversified developmental areas; economic and industry, transport and communication, space satellite, computer and information systems, genetic engineering and biotechnology and the world has become virtually a global village.

In Ends and *Means,* Aldous Huxley wrote, "Men of science have come to realize that scientific picture of the world is a partial one, the product of their special competence in mathematics and their special incompetence to deal systematically with aesthetic and moral values, religious experiences and intuitions of significance." It is the recognition of limitations of science to deal with qualitative dimensions of life that now scientists are extending the new frontiers of awareness and investigating the ground from which all life springs with new paradigms and new methodology. Julian Huxley wrote, "The evolutionary process, in the person of mankind, has for the first time become conscious of itself. We are realizing that we need a global evolutionary policy, to which we shall have to adjust our economic and social and national policies."

The discoveries of modern physics are as under:

(a) Physicists have come to see that all their theories of natural phenomena, including the 'laws' they describe, are creations of the human mind; properties of our conceptual map of reality, rather than of reality itself.

(b) Relativity theory has shown that space is not three-dimensional and time is not a separate entity. Both are intimately and inseparably connected and form a four-dimensional continuum which is called "space-time".

(c) Charles Sherrington wrote, "The material world has only been constructed at the price of taking the self, that is mind, out of it, removing it; mind is not part of it; obviously, therefore, it can neither act on it nor be acted by any of its parts".

(d) Erwin Schrodinger wrote, "All scientific knowledge is based on sense perception and that none the less the scientific views of natural processes formed in this way lack all sensual qualities, and therefore cannot account for the latter".

Creativity: Science – as defined by Hideki Yukawa – "is a manifestation of man's efforts to open up the unknown world that lies before him, to discover new possibilities .... It is a discovery of a truth in nature hitherto unknown to man". A scientific mind is precise, clear, unprejudiced, which does not condemn, which observes and sees; so it moves from fact to fact, and deals with the *present* and *now.* Creativity implies discovering something that nobody has hitherto known about or inventing something new. "Fundamental discoveries which go deep into the nature of things are the result of uncommon intuition and faith, not cold logic". The secret hidden or latent in Nature releases Itself; to Itself, or in other words, the latent manifests Itself. All *intuition, invention, discovery* are expressions of the Unmanifest; and some scientists-philosophers-mystics have called this *"triple truth"* as Intelligence or Wisdom.

Intelligence is an unconditioned energy born of the emptiness of *silence.* It is not a cerebral movement any more, it has nothing to do with the "me", the "I-consciousness", with the physical or psychological inheritance. Intelligence expresses Itself in a pure mind-brain-body system.

Innovation : The moment you think that "I want to use the *intuition* for commercial activity", the quantum energy potential falls to the materialistic and dualistic level. This level is referred as the psychological mode of mind, or ego-structure. This forms the roots of intellection and is therefore next to impossible to uproot by intellection. This demands a rigorous, consistent and persistent technology of"objective observing and understanding". So the innovation refers to an idea of converting the ego-structure to commercially useful process/ product. Man has been making efforts to achieve innovation for uncountable eons. Basically, he has been dealing with the "outer fringes" of the challenge. He has failed totally to touch the deepest level of human psyche.

Innovation is defined as a movement of"quantum energy potential of consciousness" from an unfinished process/product to a commercially useful product – as the final product is at a higher order [consciousness] than the unfinished product.

The "objective observing and understanding" leads to quantum jump of consciousness from dualistic frame of mind spectrum to non-dual and non-conceptual frame of Mind Continuum. The education of the mind-brain-body system leads to quantum jump from information [knowledge] and experience to understanding; from dualism to wholeness of life; from reason to intuition; and from intellect to Intelligence. The Intelligence resolves all challenges of life. The problem facing mankind is to become aware with a new dimension of consciousness which will generate a new relationship between man and nature, man and man, and man and environment.

The Ego is a Myth: Violence is a cerebral, neurological and chemical way of behaviour. We have been conditioned to behave violently as a reaction to certain challenges – say anger, fear, jealously, reactions and security. The content of thought has been given to me, fed into the brain by the human race. The content of thought when it changes from physical to psychological is called the ego, the me. The ego has not got any objective reality, it is a myth. The cerebral movement taking place in the body having the content conditioned by humanity has no separate point, ego self or me. It is something collective, and is expressed individually.

The ego has no factual reality at all. It has a conceptual reality. It has an ideational reality and it has a relevance to our lives. The thought-ego system cannot be separated. When the unconditioned energy of *"intuition"* encounters the conditioned mind-brain-body system [psychological mode of mind], the dualism, and "thought-ego" coupled with psychological imbalances drops away. It loses materialism [dualistic thought] and takes over to intuition.

Creativity: The creativity is at the highest quantum energy potential of Consciousness, which is invariably at the non-dual and non-conceptual frame of reference. The innovation is at the lowest quantum energy potential of Consciousness, which is invariably in the dualistic frame of reference; it is a materialistic process, it is a business activity. The dualistic thought-ego knocks down the innovation process.

There is a questioning today, "Has humanity taken a wrong turn which has created this chaos in the world? And what needs to be done within the consciousness of man to preserve our very existence and bring out a new culture?" We have to look towards "objective observing and understanding".

The answer to the human paradox lies in ending the thought-ego system through non-action of the mind, effortlessness of the mind. It is "objective observing and understanding" which is at a higher hierarchical potential of consciousness that can give *objective fact* which rejuvenates the innovation process. The survival of the human race is embedded in our faculties of creativity and innovation.

The aim of this research paper is to initiate an open dialogue with freedom and non-authoritarian way between scientists-philosophers-mystics. When innovation comes in direct encounter with the intense energy of insight [awareness], the challenge of innovation is resolved and it gives "objective truth". We intuit that the R & D potential of production is bound to enhance three times if there is deep observation and understanding of the whole creativity and innovation.

Your consciousness is shared by all humanity, so the fact is shared by mankind. The expression of consciousness of "Creativity and Innovation" is not limited to the DRDO scientists but broadly shared by mankind.

Mankind needs a higher level of consciousness, a quantum jump in consciousness – from knowledge to understanding, experience to perception, reason to intuition, and intellect to Intelligence. George Wald, Nobel Prize in

Medicine, 1967, wrote, "All the mystics through the ages have agreed that to achieve wisdom one must quit the brain, stop using it as a tool, put aside knowledge and learning, try indeed to free one's consciousness, empty one's mind, and in quiet see what enters". It appears that we have to transcend intellect and reason and imbibe intuition. We need absolute discontinuity of conditioned mind and this bestows a quiet mind. A quiet mind can save the human race from the brink of self-annihilation, it has got to find out an alternative way when we achieve higher perception.

There are many eminent scientists-mystics who talked about radical transformation in the "vertical dimension" of consciousness. Both Teilhard de Chardin and Sri Aurobindo foresee the coming of a new race and consciousness – the Noosphere and the Supermind. Julian Huxley thinks that the great opportunity today will be found in applying scientific methods in the exploration of man's inner life. Recently Eckhart Tolle wrote, "An unprecedented shift in consciousness in indeed happening on our planet". He added, "you are here to enable the divine purpose of the universe to unfold". We intuit a creative manifestation of life on planet Earth.

Discussions of Eminent Statesmen/Leaders

Recently, many eminent statesmen/leaders have also discussed "Creativity and Innovation" as under:-

(a) President Dr. A.P.J. Abdul Kalam addressed a seminar at the University of Delhi on December 31, 2004 and discussed "creative leadership". In another seminar at IIT, Delhi on March 31, 2005 he said that India should "become a continuous innovator and creator of science and technology – intensive products".

(b) R. Chidambaram also advocated, "Generation and Innovation and also support system". M. M. Sharma, on July 14, 2004 said, "We as a nation will have to innovate to survive".

(c) Prime Minister Dr. Manmohan Singh has constituted a "National Knowledge Commission" vide Gazette Notification NO. A 43011/17/2005-Adm 1 of June 13, 2005 to sharpen India's knowledge edge".

(d) Prime Minister on August 3, 2005 said, "On an average, more than 3,50,000 engineers and 5,000 Ph.D. scholars graduate from our universities and colleges every year". He added. "We cannot afford to lag behind the rest of the world". With this manpower available, this exploration is bound to enhance R&D potential [about 3 times]. We intuit that if India wants to become a developed nation in less than a decade, our statesmen/leaders, our political parties, the media, and scientistis come forward here and now with affirmative action towards innovation on behalf of the whole human race.

(e) Prime Minister, inaugurated of the 94th session of India Science Congress, The HINDU, January 4, 2007, PM said, "He urged intellectuals and scientists to give suggestions to incorporate science research and scientific temper". It is destiny that we have failed so far to regenerate the human mind-brain-body system.

One is deeply struck that man in spite of millions of years of reason, experience and information [knowledge] has utterly failed to end the conditioned psyche. There is a large gap between the superhuman task and the frailties of human endeavours, even if they are done with the best of intentions. We have to take part in the dialogue. We request other scientists to respond to this unique call. Perhaps the dialogue may lead to a new answer to this paradox.

We have to jettison reason and intellect; the experience and information [knowledge]; and imbibe intuition. The ***intuition, invention*** and ***discovery*** belong to the direct expressions of nature. Nature, in its benediction, is full of harmony and order, non-violence and non-conflict, love and compassion, perfect sustainable freedom and peace. A man with right "observing and understanding" is free from political activity, religious conformity and tradition. It implies that he is free from the weight of time, free from the burden of the past, free from all the actions of will. This only is the social psychological and even the political revolution.

The ultimate declarations that nature is discontinuous in its core, more like a staircase than a ramp... You have the ultimate unit doing the dance all by itself — that is the beauty and elegance of it.... You see the trapped ion from blinking on and off, and each blink is a quantum jump. It is a striking illustration that things occur discontinuously in nature.

If one step of staircase is at a higher quantum jump of consciousness, then the next lower step at a lower quantum jump is knocked off. This is called psychological mutation. "Objective observing and understanding" is embedded in a non-dual and non-conceptual frame of reference of Mind Continuum, and the false instantly drops away and gives the truth.

Life in homogenous, undifferentiated and non-dimensional wholesome Eternal life is an expression of Eternal intelligence.

11

Creativity and Innovation

This booklet entitled *"Creativity and Innovation"* ushers us into new millennium and new century. It is leading us to competitiveness in the global business environment. Rapid pace of globalization and technical progress have directed every economic process to competitiveness.

1. Historical: Creativity and Innovation are known for uncountable aeons. Hideki Yukawa wrote, "Man has things in him that are even unknown even to himself. He still retains within him many things that he possessed before man became man. He has them whether he realizes it or not. The human emotions are closely related to these... Thus, human emotions, and human happiness in turn, are closely bound upon with these indefinable:. The psychological debris in man is his past psychological personality.

Hideki Yukawa added, "In ancient Greece, there not only were intuition and abstraction in complete harmony and balance with each other.... " All great saints and sages have highlighted that we must transcend intellect and imbibe intuition. To sum up, creativity is intuition and transcending psychological is innovation through "objective observing and understanding".

Creativity and innovation are essential to business success as there are two pillars on which the progress of man rests. President A.P.J. Abdul Kalam addressing a seminar at the University of Delhi on December 31, 2004 discussed "creative leadership". In another seminar at IIT, Delhi, on March 31, 2005 he said that India should "become a continuous innovator and creator of science and technology - intensive products". R. Chidambaram also advocated, on November 15, 2001, "Generation of innovations and also support systems". M.M. Sharma on July 14, 2004 said, "We as a nation will have to innovate to survive". We intuitively perceive that the survival and excellence of man is embedded in "self-education", which flows to discovery and which Itself guides for the mutation

The innovation was reviewed by THE HINDU on Nov. 8, 2006, Nov. 17, 2006, Nov. 20, 2006 and Nov. 25, 2006. The 20th National Conference on "Enhancing India's Manufacturing Competitiveness" was held on Nov. 16, 2006; and 62nd "All India Textile Conference" was held on Nov. 24, 2006. But in spite of colossal effort, we have touched the "outer fringes" of innovation problem.

The aim of this booklet is to initiate an open dialogue with freedom and non-authoritarian way between scientists-philosophers-mystics of the world.

2. Creativity: It implies discovering something, which nobody has hitherto known about, or inventing something new. It explores the *intuition*, the *invention,* the *discovery* of a truth in Nature hitherto unknown to man. It is Nature's response to man's challenge, it is a pure or direct perception of Nature. Subjective science is the exploration of truth in Nature and in man.

3. Innovation: It refers to the use of Nature's response at a lower quantum energy potential of awareness which gets transferred to a commercially useful process/product. It is not a functional activity, it is a business activity, a materialistic process, man-made structure. Competitiveness in the global business environment is characteristic of 21st Century. Rapid pace of globalisation and technological progresses lead to innovation: It leads to manufacturing, marketing, sales, profit. Innovation uses the basic principle of awareness for transformation.

Innovation is the *violence* of the conditioned mind-brain-body system., as all psychological imbalances leads to violence. One of the main reasons is that the *dualistic* approach and *thought-ego* coupled with the psychological imbalances have been so pernicious is that the error of dualism forms the root of intellection and is therefore next to impossible to uproot by intellection. Innovation can never be resolved through reason, intellect, ideals, and concepts, thought structures.

4. Man-Made Universe: The world extended in space-time-causation matrices is a construct of our conditioned mind-brain-body system. It is our representation. All psychological imbalances – violence, conflict, fear, greed, anger, terrorism, and so on are constructs of the waking state of mind or relative frame of reference of mind spectrum. The socio-economic-political systems, the institutionalised religions, the reforms in the external field have so far failed to transform man. Violence is a construct of symbolic-dualistic frame of reference of mind spectrum, and nonviolence is the percept in the non-dual and non-conceptual frame of reference of Mind Continuum.

"I-consciousness" has no factual reality at all. It has a conceptual reality. It has an ideational reality. It exists coupled with thought. At the psychological level, the mind cuts itself into subject and object, and there is transference of energy between the subject and the object; the object [memory] gets

Fact [Subject] + Recalled Memory ⟶ Mutilated Fact + Modified Memory modified and gets recorded; the subject also undergoes change. The interaction, in the language of physics, may be discussed as:

$$\text{Bare Cognition} + \text{Conative} \longrightarrow \text{Mutilated Cognition} + \text{Modified Conative}$$

In the language of transpersonal psychology or deep ecology, we call pure perception or fact as 'bare cognition', and recalled memory as 'conative'; so we write this as :-

The past is the thought-ego; there is no thought-ego in the present. All innovation is governed by the thought-ego, which is always the past.

Thought is the response of memory which is the past. All continuity is thought; where there is continuity there is nothing new. The past and the future can only be thought of as the present, 'past' must be identified with memory and 'future' with expectation; memory and expectation being both present facts.

There is a singular failure of the human mind to resolve our challenges, our conflicts, our violence. The answer to this paradox is given by Albert Einstein in prophetic words, "Objective observing and understanding". Observation is possible in the reaction-free attentiveness. Observation and understanding are in the non-dual realm.

5. Ending of Thought-Ego: To end violence demands a rigorous, consistent and persistent technology – "objective observing and understanding" – capable of pursuing dualism and thought-ego to their limits.

The moment the desire [the ideal, the motive] is born to use the *intuition or invention* or *discovering* for commerce, for materialistic pursuit, it is bombarded by "thought-ego" which destroys the innovation. Each step there is a movement away from the false in our daily life, in our daily relationship. The enquired has to expose himself to encounter the unconditional energy and the awareness releases the "objective truth". The false drops itself, and achieves that innovation. It suggests that it is the awareness that heals. The awakening of this perception is born of the non-duality of wholeness. This lays a new foundation for innovation, a new relationship of man and nature, man and man, and man and environments. The enquiry is not a commitment of the intellect, but it is commitment to whole of life, the total creativeness.

One has to be aware from the beginning to the ending, not to become inattentive in the middle of it. The new quality of awareness is attention, and in this attention there is no frontier made by thought-ego. This attention is also called supreme Intelligence. It is the Intelligence, which resolves all challenges.

6. Limitations of Conditioned Psyche: No human problem, no problem of human relationship [including innovation] can ever be resolved through knowledge, through thought-ego, on the cerebral level. Man has been trying to resolve the challenges of innovation for uncountable aeons, but has utterly failed, as he has been dealing with *"outer fringes"* of conditioned psyche. Knowledge, experience, intellect, memory do not result in direct perception. All human challenges go to abeyance when the subject of cognizance attains a higher level of awareness.

All great discoveries have arisen from the realm of the timelessness; so scientific enquiry is not a cerebral activity and gathering knowledge [information] but an expression or a quest of the Timelessness, the Divinity, the Life Field. Innovation is to use the discovery to a commercially useful process/product. Discovery is fundamental and innovation is a derivative. Innovation is embedded in the depth psychological dimension of man but the tool to be followed is scientific research, or scientific temper.

7. Concluding Comment: We intuit that (i) if India wants to become a developed nation in less than a decade, our statesmen/leaders, our political parties, and the media, and our scientists come forward *here* and *now* with affirmative action towards innovation on behalf of the whole human race; (ii) our publication industry may take up on priority to publish books on "Symbiosis of Science and Spirituality" and allied explorations, say, "Creativity and Innovation; (iii) the "true aim" of education is the non-dual and non-conceptual mode of perception; or education, in the true sense, is understanding of oneself, for it is within each one of us that the whole of existence is gathered; (iv) the knowledge, experience, intellect, memory do not result in direct perception, as all human challenges go to abeyance when the subject of cognizance attains a higher level of awareness, and (v) the survival of human race is embedded in our faculty of intuition [insight] and insight self-express as itself as ending of "innovation".

Intellect is "thought-ego" constructing dualism, violence and conflict. Intelligence [Wisdom] is the capacity to feel as well as to reason, it is invariably in the non-dualistic realm, and it is beyond mind-brain-body system, or non-sensual direct perception. Until we approach life with Intelligence, instead of intellect alone, no political-economic-social or educational system is the world can save us from the toils of chaos and destruction.

12

Science Research and a Scientific Temper

1. Introduction

The Constitution of India, Fundamental Duties of Citizens, Chapter IV A, Article 51 A(h) reads as, "To develop the scientific temper, humanism and the spirit of enquiry and reform". Recently Goverdhan Mehta on July 18, 2000 said, "We have to identity science with creativity".[1] Again, Jayant Vishnu Narlikar on April 9, 2001 said, "What is needed is a campaign for scientific temper".[2] A Mind with a scientific temper implies to be free from all thought-structures in the intense energy of insight. It is tuned to perceive the Wholeness of Life, which bestows a perception of intuition [insight], spontaneity, nonviolence, harmony, human values, truth, love and compassion, so it is in communion with the Universal Consciousness, the Life Field.

The Prime Minister of India has appointed *"The National Knowledge Commission"* vide Gazette Notification No A 43011/17/2005 –Adm 1 dated 13 June 2005. As all knowledge is rooted in the Universal Consciousness or Life Field, so it is expected that the National Knowledge Commission will recommend the implementation of higher awareness or consciousness at the National Level.

This article explores a technology (not methodology) to have a new approach to the art of living. The last 10,000 years, or more, have demonstrated that the task of turning the age-old tide of human violence, aggression, ambition and motive is a stupendous task. Science also brought man at the control panel of human mind-brain-body system. We are deeply struck by man's concern about the large gap between the superhuman task [to live in peace] and the frailties of human endeavors. We intuitively observe that Nature is ready to help mankind within the scope of intuitive living. We have to keep in mind that intuition is fundamental and rationality is a derivative. This article advocates transcending rationality and explores the scientific temper, which blossoms to innovation in science and in man. It also discusses education, meditation, discovery and innovation, creative leadership, and new relationship of man with nature.

2. Intellect, Reality and Truth

J. Krishnamurti, an eminent philosopher-mystic, talked of Truth, Reality and Actuality; and these are discussed as under:

Actuality: It is what the senses cognize, but the senses are extremely limited in their perceptive capacity. Actuality is confined to the frontiers of the sense knowledge. The senses know only a fragment and the whole is not a mere extension of the fragment.

Intellect: Reason or rationalism is a precious value thrown up by evolution and the source of much human progress in culture and civilization. It is always conditioned, so it is incapable of seeing truth. The intellect, or relative frame of reference, or conditioned mind is at the lowest quantum energy potential of consciousness. Mind is inclusive of all identifications and measurements, yet a dimensional reality of every psychological level represents an illusory deviation.

Reality [Psychological Mode]: It is what thought of man has put together; it is man-made cosmos. Mind [with a lower case 'm'] is dualistic and repetitive. There is an interaction between the subject [past] and the object [present]. This leads to coherent superpositions of different webs of recalled memory [past] and the fact [present]. Reality is in the field of mind's measurement. Thought is response from memory, which is reactive in nature. This leads to distorted perception or conditioning. Thought-structure is repetitive and mechanistic. It has no capacity for newness or originality. Thought is sequential, successive, one-dimensional, while the real world presents itself as multi-dimensional, non-successive, simultaneous patterns of infinite richness and variety.

Reality depends on one's scale of observation – which is the level of awareness of the subject of cognizance. In fact, the observer and his scale of observation are not two different things. The scale of observation is the observer.

Reality [Scientific Temper]: Discoveries come as intuitive flashes. Discoveries come in a non-dual and non-conceptual frame of reference, or it is timeless perception or true perception. Scientists observe a non-dual thought which expresses itself in scientist's brain-body system, so it is not a brain or sensual activity, but nature uses the brain as a media. What we call Mind [with a upper case 'M'] is a non-dual frame of reference of Mind Continuum. So *creativity, discovery, intuition, insight,* is in science [observation] and is hidden in man.

Einstein said once that the most incomprehensible fact is the fact that nature is comprehensible. Creativity and intuition deal with this miracle – through which human thinking is capable of penetrating into the inner working of nature. One aspect of the miracle is that nature is equally comprehensible to any scientifically trained human being, whatever his or her origin or cultural background.[3] Nature invariably responds to experimental interrogation.

In objective science, the subject of cognizance is at the immanent quantum energy potential. The time-space matrices (objective world) is at the successive consciousness. There is no energy transfer between the object and the subject. The objective scientific mind is very factual, so the subject sees the object from *fact* to *fact*. It is a movement from the *known* to the modified *known,* which has higher sensitivity – it pursues materialism. The objective science constructs new models of known artifacts with a higher sensitivity. It leads to invention, which flows to a new model of aeroplane, air-conditioner, and so on. It shows that the achievements of objective science are limited – it adds to human comfort, leisure and pleasure. It also contributes to competition, motivation, violence, disorder and strife. It constructs weapons of mass destruction [WMD].

Truth: It is truth which is the Whole; it exists by Itself. Truth is not cumulative; it is from moment to moment. It is the highest quantum energy potential of Consciousness, or the highest frequency of Consciousness. Love is not different from truth. Love is that state in which the thought process, as time, has completely ceased.

Albert Einstein, in a prophetic vision, stated that we need, "objective observing and understanding", which will lead us to "peace and serenity". We opine that the Actuality, Reason and Intellect, and Reality [Psychological Mode] are sterile in bringing Innovation. Unless we put order in the world of reality, we cannot make any progress. You cannot impose order on top of disorder. You cannot impose nonviolence on a consciousness which is violent because that violence has deep-rooted causes in the psyche. We need scientific temper, or non-dual and non-conceptual Mind to perceive Innovation. Innovation flows from discovery.

3. Imbibing Transcendence of the "Objective" and Imbibing the "Subjective"

(a) Krishnamurti said, "India has lived all these centuries on ideas. She has to move from a life based on ideas, which are non-facts, to seeing of fact only".[3] We have to jettison living in the *non-fact,* the *false,* the *past,* the *what should be!,* and learn to live in the *fact,* the *true,* the *present,* the *what is?* Krishnamurti added that we need, "a way of perception, of challenge and response from a state that lay beyond the mind, beyond brain, memory, all responses of consciousness".[4] We opine that the mind is transcended in the objective field and goes to the inner realm [subjective realm]. Ilya Prigogine and Isabelle Stengers stated that, "the observer is situated in the observed world. Our dialogue with nature will be successful only if it is carried on from within nature.[5] The new vision on education is to transcend the objective and embrace the subjective. The subjective is the new dimension, which we include in the curriculum of education.

(b) No education worthy of the name can be conceived as something external having no relationship "with man's innermost being and becoming", and task of education is inward transformation or inward flowering, and character building. Education is to understand the psychodynamics and limitations of the conditioned psyche and to understand the Unconditioned Energy. The foundation of education is not knowledge [outer] but intuition [inner].

(c) Education may mean opening the doors of perception on to the vast movement of life. It implies the understanding of life directly, instead of in the abstract, linear terms of representational thinking. It bestows awareness of the fact, insight [intuition], spontaneity and right action. This guarantees nonviolence, peace and human survival.

(d) The answer to mutation of the conditioned psyche is embedded in Albert Einstein's prophetic words, "Objective Observing and Understanding". Max Planck, Nobel laureate in physics, called it a single point of our "individual awareness". Edwin Hubble called it "world of pure values" which flows from insight [intuition].

(e) All perennial knowledge is rooted in the Universal Consciousness, or Life Field. All great discoveries have arisen from the realm of the timelessness, so scientific enquiry is not a cerebral activity and gathering knowledge (information), but a quest of the Timelessness, the Divinity. Discovery is not activity of the brain or the senses, but nature uses the brain as a media. The foundation of education is higher awareness or consciousness. It is an enquiry in utter freedom in a non-authoritarian way. And this bestows higher awareness. There must be a state of mind in which there is "direct perception", or awareness of the wholeness and oneness of life. This flows to creative leadership. We have to introduce holistic education, which implies to meet life in the timeless *here* and *now*. It fulfils the aspirations of mankind – Order, Harmony, Nonviolence, Peace and Bliss.

New Paradigm in Education in UK: The Scientific and Medical Network, UK, which has over two and a half thousand members, have started a New Educational Initiative, called the Blaker Foundation. "It aims at a transition from our evolving but materialistically based culture to a broader spiritually inspired understanding of the world". Blaker put it succinctly, "Our science of today cannot stop the spiritual advance, but can only delay it for a while, for such delays will automatically drag in their wake calamitous consequences, including geological disasters". There is a silent renaissance goi g on in UK and Europe. In some universities, UK has started conferring M.A., M. Phil., and Ph. D. degrees. It is exploring a new educational structure "to allow teaching within a spiritual and holistic context.

4. Mutation/Transformation

J. Krishnamurti wrote, "There must be a radical revolution. The world crisis demands it. Our lives demands it ... there must be a fundamental, radical revolution, because everything about us has collapsed. Though seemingly there is order, in fact there is slow decay, destruction: the wave of destruction is constantly overtaking the wave of life". He added, " Seeing the false as the false and the true as the true is transformation, because when you see something very clearly as the truth, the truth liberates .... Truth is not cumulative.... Eternity is in the now."

It is not the repetitions of the word, nor experiencing of a vision, nor the cultivation of silence. The bead and the word do quieten the chattering mind, but this is a form of self-hypnosis. You might as well take a pill.

Understanding demands self-knowing; learning about oneself is endless, and the beauty and the greatness of it is endless. But self-knowing is from moment to moment, the self-knowing is only in the active present; it has no continuity as knowledge.[6] Anything that is cultivated is a product of thought. Similarly, anything that the mind recognizes is also a product of thought. Transformation is freedom from all choices, or it is a state of Choicelessness.

It is the Intelligence which conveys to the world of reality the whisper of Truth. The intelligence is born in the interval between challenge and response, in that moment where crisis remains unsesolved. Meditation is a never-ending movement. Meditation is emptying the mind of the known. If a mind with a higher quantum energy potential of consciousness bombards the *known,* it dissipates the *known.*

Mutation means radical qualitative change in the totality of our consciousness, it is a change in the wholeness of the mutant. Transformation is not in the future. It can only be *now*, from moment to moment. Transformation is seeing the false as the false and the true as the true. It is the truth that liberates.

The act of "observing, listening, understanding" is possible when there is attention, and attention does not exist if there is interpretation, evaluation, condemnation and identification. The solution of the problem lies in the problem itself. If a seed of truth is planted, it must operate, it must grow, it must function, it has a life of its own.

5. Discovery and Innovation

Hideki Yukawa, a Japanese Nobel Laureate in physics, wrote, "Man discovered new things hidden in Nature .... discovery of a truth in nature hitherto *unknown* to man .... Science is a manifestation of man's efforts to open up the *unknown* world that lies before him, to discover new possibilities .... A discovery may lead to happiness and prosperity, or it may, equally well, lead to a loss of humanity and the destruction of mankind".[7]

Discovery demands a rigorous, consistent and persistent scientific research by scientists. Discovery is fundamental and innovation is a derivative. Innovation is embedded in the dimension of man but the tool to be followed is objective scientific research. The innovation is in the material field, but it emerges from Discovery, a non-dualistic perception. Discovery is a quantum jump from dualistic frame to non-dual frame, quantum jump happens through discontinuities and not through a linear path. Innovation [technology] is the systematic orchestration of all knowledge and experience to lead to something practical and commercially useful.

Discovery or invention is a new idea that works and Innovation refers to new idea being converted to a commercially useful process/product. It is not a functional activity; it is a business activity and you need every component of business activity – manufacturing, marketing, sales – as a part of it. It is a materialistic process, so "I-ness" or ego enters the whole process spuriously with its damaging effect. This reduces the perception to a concept and leads to distorted perception. This hinders the progress of the innovation process/product.

When thought travels in the body, it consumes nervous energy. When an emotion travels in the body, it affects the chemical system. In the timeless realm, the nervous tension and chemical pressure go to abeyance, and the whole system is relaxed. It is a balanced and equipoise state. Discovery, intuition, insight, creativity are in science and is hidden in man. Creativity and intuition are faculties through which human thinking is capable of penetrating into the inner working of nature. Nature invariably responds to "experimental interrogation", when the whole system is governed by Intelligence.

We have to jettison experience and knowledge [information] and imbibe intuition and spontaneity. Man has to learn to jettison experience and knowledge and to culture direct perception.

6. Creative Leadership

When one is aware of the objective truth, and ignorant of the subjective world, the Mind is engaged in invention. If one is simultaneously aware of the subjective world [inner] and the objective world [outer], then the Mind is engaged in discovery. Awareness of the subjective realm leads to creative leadership. The higher the potential of creative leaders in a nation, the higher the potential of consciousness of that country.

"Science with creativity" is direct expression of Cosmic Consciousness in the pure Mind. The art of living emerges from direct perception or ultimate perception; and this is free from any psychological imbalances, conflict, chaos. The art of "science [restructured science, education, ethics, health] with creativity" leads to an expression of the wholeness of life.

[restructured science, education, ethics, health] with creativity" leads to an expression of the wholeness of life.

"Science with creativity" is an experiment of creating a "New Man" – a new human being who lives in harmony with the inner and the outer, with himself and his environment, and who is free from all ideologies and conditionings that now divide humanity. The human survival is embedded in this gigantic experiment. And if man escapes his responsibility, he does so at his own peril, of India and mankind.

Some scientists have advocated that proposed convergence of Nano, Bio and Information technologies will lead to creative leadership. They are under an illusion, as creativity is only connected to education of mind. Another illusion is that intellectual knowledge will lead to alleviation of poverty. We urgently need educating the human mind so that we make a quantum jump from symbolic frame to non-symbolic frame, from knowledge to intuition [insight]; and insight flowers to innovation. We urgently need innovation for ending of our poverty.

To remain free of "ego" in the pursuit of materialism [innovation] is a superhuman task, and the "ego" acts as hindrance to innovation process/product. The progress for agriculture and industrial development is a function of "creative leadership", which is a function of state of mind. We want a non-dual and non-conceptual and timeless mind, so that it is always pliable and ever ready to answer our challenges in the *present* and *now*. Creative leadership implies communion without fear, freedom to understand each other, to communicate directly. Creative leadership implies direct perception, *here* and *now*.

7. New Relationship of Man with Nature

Common Waking State of Man: The waking state of mind, or the relative frame of reference, or the psychological mode of mind, both the subject and the object are at the same quantum energy potential, hence there is no transference of energy. In the psychological mode, there is no relationship between the subject and the object, or with people and with ideas. It constructs chaos, conflict, violence and fear. It is lack of understanding that causes conflict.

Scientific Mode of Mind: It is governed by "the principle of objectivation",[8] and the mind [object] moves from *fact* to *fact*. A scientific mind is precise, clear, unprejudiced, which does not condemn, or identify, or compare, or measure; which observes, which sees. The frame of reference is non-dual and non-conceptual, so it engages itself in direct perception or pure-perception. One lives from fact to modified fact, an order. The mind is engaged in intuition [insight].

Objective science is a dialogue with nature, it sees the challenge and responds to experimental interrogation as a response, and this is called intuition. Silence is born at the moments between challenge and response.

Objective science is a pursuit of materialism. It is an open-minded enquiry and leads to abstraction. It guides to construct new and more sensitive artifacts. It also leads to violence. Maurice H. Wilkins wrote, "The whole question of moral, spiritual, and other dimensions is normally pushed out of science".[9]

Subjective Science: When non-dual state of Mind is presented to Pure Consciousness, the Mind receives the energy from Consciousness and it makes a quantum jump from the outer field to the inner realm. The Pure Consciousness becomes self-referral, or it sees Itself. All violence and conflict go to abeyance. This is called Subjective Science, or Absolute Subjectivity, or "Superluminal [Faster-than-Light] Principle". There is order, harmony, nonviolence and peace.

State of Mind as a Function of Quantum Energy Potential of Consciousness: The evolutionary rise of life as a function of the quantum energy potential of Consciousness is shown in Fig. 1: (a) it is called the common waking state of consciousness; (b) psychological state of mind; (c) scientific mode of mind; (d) when the scientists decide to innovate, the quantum energy potential of consciousness falls due to intervention of ego; (e) the potential of consciousness again rises to scientific mode; (f) the potential rises to intuition; (g) it blossoms to invention; (h) invention rises to discovery; and (i) all violence, conflict ends and truth, love and compassion express themselves.

Man does not have anything, which is his personal – including brain and memory, which is of humanity and the timeless. Everything is a grace of nature. We are only dancing dolls where someone else [Cosmic Consciousness, Life Field] pulls the strings; we are toys in the hands of destiny. A living in communion with nature is free from all psychological imbalances, ecological crisis, strife, conflict, and so on. This is the consummation of human life.

8. Concluding Comment

The self-knowledge of mind-brain-body system guides us that the Actuality, Reason and Intellect, and Reality [Psychological Mode] are sterile in bringing Innovation. We need the non-dual and non-conceptual frame of reference of Mind Continuum, or scientific temper to perceive Innovation.

Albert Einstein, in a prophetic vision, wrote that we need "objective observing and understanding" and this leads to "peace and serenity". If the observation is continued for a long chronological time, the Mind makes a quantum jump from "outer field" to the "inner realm", and it goes to abeyance. The cosmic Consciousness expresses Itself as "Subjective Research" and there is perfect order, harmony, nonviolence and peace.

We have to educate our physical organism, and psychological structure to move into non-action gracefully, and then move to Silence. This will flow to intuition, invention and

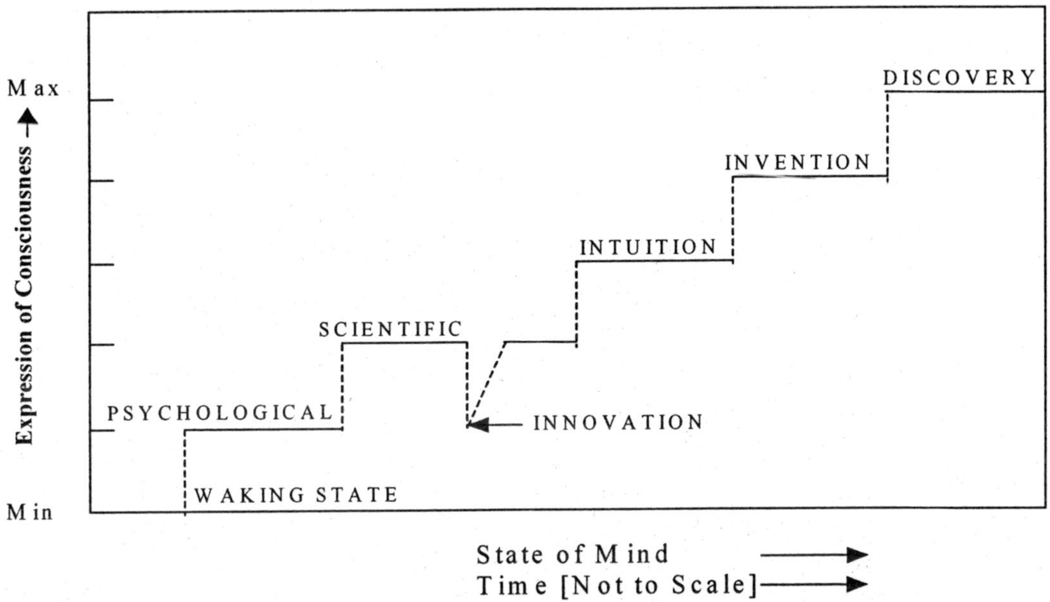

Fig. 1: States of Mind as a function of Quantum Energy Potential of Consciousness: (a) Common waking state of mind; (b) Psychological State; (c) Scientific Mind (d) Innovation [Potential falls due to Ego]; (e) Back to Non-Dual State; (f) Intuition [Insight]; (g) Invention; (h) Discovery and (i) Superluminal State [Spirituality].

discovery, which express itself as Innovation. This happening leads to Cosmic Intelligence, perfect order, nonviolence and peace.

Many eminent scientists – including Albert Einstein, Erwin Sehrodinger, Werner Heisenberg, Arthur Eddington – have stumbled upon another mode of knowing, called direct, or intuition, or insight. It is indeed direct understanding of all human psychological imbalances. It leads to discovery which has the potential to express itself as innovation.

If India is to become a developed nation in less than a decade, our statesmen/leaders, our political parties, the media and scientists must come forward *here* and *now* with affirmative action towards innovation on behalf of the whole human race. Fundamental discoveries, which go deep into the nature of things, are the result of uncommon intuition and faith, and not cold logic. And intuition, inventions and discoveries flow to innovation. Sir Aurobindo prophesied that "the initiative can come from India and although the scope must be universal, the central movement must be hers".

References

1. We circulated an article entitled, *"Innovation in Science and in Man"*, on August 15, 2005. The present article entitled *"Science Research and a Scientific Temper"* is an extension of the same.
2. Goverdhan Mehta, in *Global Education and Human Survival,* New Delhi: Sanbun Publishers, March 2002, p. 22, 23 & Jayant Vishnu Narlikar, *ibid.,* pp. 22, 23.
3. Hideki Yukawa, *Creativity and Intuition,* Tokyo: Kodansha International Ltd., 1973, Cover Jacket.
4. J. Krishnamurti, in Pupul Jayakar, *Krishnamurti: A Biography,* San Francisco: Harper and Row, 1986, pp. 379, 380.
5. Ilya Prigoine and Isabelle Stengers, *Order Out of Chaos,* Bantam Books, 1984, p. 218.

6. J. Krishnamurti, in Fritz Wilhelm, *Within The Mind: On J. Krishnamurti,* KFI, 1982, p. 123.
7. Hideki Yukawa, *op. cit.,* pp 112, 113, 205.
8. Erwin Schrodinger, *What is Life? & Mind and Matter,* Cambridge University Press, 1967, p. 131.
9. Maurice H. Wilkins, *Synthesis of Science and Religion,* Mumbai: The Bhaktivedanta Institute, June 1988, p. 30.

13

Cause of Our Failure in Innovation

1. Introduction

Science is a way of communication, a dialogue with nature. Present-day research leads us further and further away from the opposition between man and the natural world; or in other words, instead of rupture and opposition, there is the growing coherence of our perceptions and nature. Nature responds to experimental interrogation. Hideki Yukawa referred to Galileo Galilei, and wrote, "He taught us by his own achievement that experiments purposely designed are of vital importance in uncovering the *truth* hidden in nature. He asked questions addressed to nature and succeeded in receiving answers directly from nature".[1] It is a creative process, where there is unfolding the discovery of a truth in nature hitherto unknown to man.

This article discusses the cause of our failure in innovation. The cause is embedded in intuition, radical transformation through scientific temper, the ending of the ego, and reorientation of the whole educational structure. It leads to ending of violence, conflict and psychological imbalances, and bestows freedom, order, harmony, and peace for the whole mankind.

2. Cause of Our Failure in Innovation

The brain is a material thing, it has its own activity. We use a part of the brain and this is conditioned by thought. Culture is a training of the brain and the refinement of the brain; the training of the brain in action, in behaviour, in relationship, and also a process of enquiry that leads to something untouched by thought.

Psychologically, inwardly, there is only the fact, the cause. The fact is that there is no opposite except physically. There is no such thing as psychological evolution. There is nothing to learn in the psychological field. To be a light to oneself requires a great freedom, a very clear brain, not a conditioned brain.

When you listen deeply, the seed has been sown. If that fact is left alone, is operating, you do not have to do a thing about it. The fact itself is the operator, not you operating the fact. Innovation is a materialistic process, so it converts a challenge into a problem, and then carries it over and allow the conscious and subconscious to get conditioned with that burden. After all the problem is the result of our incapacity to face a challenge totally. Problems are not objective facts. They are creations that are projections of the conditioned mind.

Jealousy is nothing but, a logical consequence of competitive society. Fear is the urge for security. Frustration is just the shadow cf ambition. Interpretation is trying to understand a fact through concepts, ideas and theories. Innovation leads to summation of all psychological imbalances. The way to understand is to look [to observe] at the fact without trying to evaluate, or interpret, or without trying to judge it, without identification, naming, measurement, comparison, acquisition. Innovation sees a fragmentary perspective and action; it sows the seeds of misery and sorrow. We are victims of fragmentary approach to life.

The "ego", or the "I-ness", or the "self" or the "me" enters the whole process of innovation, spuriously, deceptively, secretly, and it damages the process through dualism. To remain free of "ego" in the pursuit of materialism [innovation] is a superhuman task, and the ego damages the innovation.

The ego has no factual reality at all, it has a conceptual reality, it has an ideational reality, it has a vibrational existence. You identify with the past events as "memory" and the future with "expectation"; and this identification is called ego. There is coherent superpositions of different webs on the object and it distorts the present perception, which hinders the progress of innovation process/product.

The ego is time; it is put together by the mind. There is no growth, no freedom, and no discovery. The movement of the mind, or the cerebral movement is absolutely irrelevant.

It is the "ego" at the materialistic level or dualistic level which is the primary cause of our failure in the field of innovation.

3. Ending the Cause Through Intuition

When the mind is confronted with something of staggering dimensions, which it cannot interpret there is nothing to refer back. The nature responds and it provides answer to that challenge, and this is called intuition. The spontaneous silence which arrives through the realization of its limitations, has the quality of dynamism and nature providing answer to the challenge has a quality of positiveness, dynamism and creativity. This creativity is called intuition.

The non-dual and non-conceptual frame of reference of Mind Continuum leads to "objective observing and understanding". Understanding of anything is a personal discovery of the content there of. The scientific frame of reference moves from *fact* to *fact.* Discovery is its mission, its perception.

The non-dualistic Mind leads to intuition. It has nothing to do with the act of volition, with the conscious mental activity.

Wisdom is the intuitive perception, intuitive discrimination of what is right, and what is wrong.

Science with creativity is direct expression of Universal Consciousness. Fundamental discoveries, which go deep into the nature of things, are the result of uncommon intuition and faith, not cold logic. Creative leadership implies communion without fear, freedom to understand each other, and communicate directly. Creative leadership implies direct perception, *here* and *now*.

Intuition is synergistic to innovation, or in other words, intuition is at a higher quantum energy potential of consciousness.

Many eminent scientists – including Albert Einstein, Werner Heisenberg, Erwin Schrodinger, Arthur Eddington – have stumbled upon another mode of knowing, called direct, or intimate, or insight, or intuition, or "the understanding of life directly, instead of in the abstract, linear terms of representational thinking". The non-variable consciousness (the objectless attention) takes over as the true subject – this is called a Mind with a scientific temper, it is tuned to perceive the *fact,* the *true,* the *present,* the *what is?* This reveals the psychodynamics and limitations of the conditioned psyche; the conditionings go to abeyance; and it bestows a perception of intuition and spontaneity. The pure perception bestows non-violence, harmony, human values, truth, love and compassion; so it is in communion with the Universal Consciousness, the Life Field.

A mind that has set aside seeking in the future, and has denied its past is ethical – this is also called a Mind with scientific temper. A non-dual and non-conceptual frame of reference is ethical. The ethical nature is beyond conflict and psychological imbalances. An ethical mind is synergistic to psychological mode of mind.

David Bohm, an eminent physicist, stated that: Observation is perception without psychological involvement, without any kind of reaction. Reaction implies evaluation, judgement, condemnation, comparison. Observation is non-evaluatory perception. Observation without the pseudo-observer is the state of wholeness of perception and wholesness of response.

4. Radical Transformation Through Scientific Temper

An enquiry by scientific temper of Mind to observe the psychodynamics and limitation of the conditioned mind-brain-body system lead to intuition which blossoms to Silence and flowers to Intelligence.

Mutation or radical-transformation is unconditional total change in the whole human consciousness – the conscious, the subconscious, the unconscious. Some scientists call the conscious as short-range memory, the subconscious as long-range memory, and the unconscious as memory that is forgotten.

During the last 10,000 to 30,000 years, we tried relationship of the mind to acquisition, accumulation and preservation of wealth; relationship to ownership, property and prestige; relationship to knowledge, experience and authority – but nothing has changed. The answer to this paradox is provided by Albert Einstein in words, "Objective observing and understanding", which leads to "peace and serenity". Observing or enquiry is seeing the conditioned psyche without any identification, measuring, and comparison, without any word or thought or time. Observation transcends all psychological imbalances.

It is to observe the whole psychodynamics and its limitations of the conditioned psyche; it is to observe in total freedom. Freedom is existing as a fact within me. It is facing the fact of our existence; it is unrelated to our mental activity.

It is related to perception and understanding. Perception is the medium for mutation.

Love is not different from truth. Love is that state in which the thought process, as time, has completely ceased. Love is a state of being in which the movement of life is unsoiled by the touch of conditioned mind. Life is moving, and the movement of life untouched by the mind is the state of love. Love is state of being in which one has arrived at the humility to act without ego. Where love is there is transformation. Freedom, innocence, maturity and love, will have to come through the complexities of the life that we are living.

5. The Ending of the Ego

The materialistic values are local, parochial, temporary, transient in application. These emerge from identification, measurement and comparison. These may change with the change of social circumstances these are external relationship of the individual with society. These are construct of the intellect. The innovation governed by ego has materialistic values which are transient and as such the whole process is not sustainable.

The innovator has to explore the innovation-technique to make another quantum jump to scientific temper and this quantum jump will bestow intangible values. Let us explore a typical case; if a friend says that the present innovated-technique is faulty, you feel insulted and you react to the insult. The fact is only those words someone has uttered. That is an objective fact. But the innovator offers subjective reactions to that fact – it is not an objective fact, but a subjective reaction. The innovator has to see his subjective reaction.[2]

The innovator should not convert subjective reaction to a problem. Instead be sees the subjective reaction with all-inclusive attention, taking in its fold objective situation, subjective reactions and the roots of the reaction – such an approach is bound to end all ego.[2]

6. Reorientation of the Whole Educational Structure

All teachings by eminent philosophers-mystics – such as Sri Aurobindo, Tagore, Zakir Hussain, Radhakrishnan, Mahatma Gandhi, and so on were formulated at the non-dual frame of reference of Mind Continuum [*Paravidya*, higher knowledge], which is beyond the capacity of the symbolic-dualistic frame of reference of mind spectrum [*Aparavidya,* lower knowledge].

It appears that for the first time in history, the scientific spirit of enquiry is being turned upon: (a) the dynamics and limitations of mind-brain-body system; (b) the dynamics of evolutionary insight; (c) modes of knowing – Aparavidya [lower knowledge, or empirical, rational and analytic mode of knowing], and *Paravidya* [higher knowledge, or mystical or intuitive modes of perceiving, or understanding of life directly]; (d) the understanding, the learning and the right action; (e) the quantum jump from symbolic-dualistic frame to non-dual frame of reference of Mind Continuum; (f) the state of awareness; and (g) the state of meditation. This has introduced a new transformative technology which flows to new relationships, art of renewal, and communion with the new source of energy (Life Field). This bestows order, harmony, freedom, non-violence, and bliss. This technology integrates the left hemisphere and right hemisphere and thus making the brain holistic. It bestows a perception of the wholeness of life. The transformative technology offers a cultural reconnaissance, a reformation of life.

Action Plan: a) The teaching must be confined to *"how to think"* and not *"what to think"*. How to think implies awareness of the mind-brain-body system, or awareness of the wholeness of life. It involves "observing, listening, understanding and living", so it is synergistic to "what to think".

(b) To educate the educator is also termed as "teacher training course". To educate the educator is to have him understand himself.

(c) It is essential to incorporate science of mind in their curricula, in their studies.

(d) To conduct courses at all levels – research scholars at doctoral level or post-graduate level especially in the science faculties of different universities and at lower levels in schools and colleges.

(e) We have to introduce psychodynamics of mind-brain-body system in science faculty and transpersonal psychology in art faculty as a part of educational curriculum at University/College level.

(f) The Action Plan has to be multifaceted and multidimensional. The fundamentals are "to educate the educator or training of teacher", and writing of books both for the teachers and students; and naturally these are at different levels, and in different languages. This is a stupendous task.

7. Meditation

Meditation is a new approach to life, in which we look at the objective challenge, the subjective reactions and the causes of the subjective reactions simultaneously, without identifying oneself to any of them. To get introduced to the intimate are delicate inter-relationship between the body and the mind is meditation.

It is an eternal voyage where we are constantly learning. We receive the message that every moment is unfolding to us. Learning has no end. Accumulation has an end. Learning is growing with life, moving with life. That is the beauty of living.[2]

8. **Concluding Comment**

Through all these centuries, we have been trying tc sublimate the mind, to discipline the mind, but we have utterly failed. The cause is that in order to see itself, the mind must cut itself up into the subject and the object; and whatever it sees is only partially itself. It implies that intellectual process is sterile in transforming the human psyche; as no observing system can observe itself observing, or the seer cannot see itself seeing.

Albert Einstein, in prophetic words stated, that we need, "Objective observing and understanding", which will lead us to "peace and serenity". Observation is to see the limitations of the psychodynamics of the conditioned psyche, this bestows awareness which blossoms to Silence. The Silence grows to Intelligence; and this leads him to grow in relationship without the intervention of the old mind. The past goes to abeyance and one lives in the *present,* the *now.*

Discovery is fundamental and innovation is a derivative. Innovation refers to an idea being converted to a commercially useful process/product. It is a materialistic process. It is to enhance the awareness of the psychological mode of mind, [relative frame of reference] to scientific temper. It is a movement from indirect knowledge/experience to direct perception or pure perception.

It is only the holistic approach from spirituality that can result in radical transformation. The discoveries blossoms to innovation. There appears to be a strong evidence that *Nature* has stirred up the pool of Universal Consciousness and this has resulted in the urgency for sustainable development of new techniques to take on global competitors through innovation.

References

1. Galileo Galilei, in Hideki Yukawa, *Creativity and Intuition,* Tokyo: Kodansha International Ltd., 1973, p. 111.
2. Vimala Thakar, *Mutation of Mind,* Ahamedabad: New Order Book Co. August 1972, pp. 35, 131, 171, 173.

Questions and Answers

Q. What is the cause of our failure in innovation?

Ans. The innovated structure is governed by thought-ego, by dualism and psychological imbalances. The innovated structure is a materialistic and an economic activity. The innovator thinks that the cerebral organ, or the activity of cerebration can resolve the challenges. He is totally dependent on intellect and reason, which works on measurement and psychological time-space matrices. In the field of innovation, the failure in discovering the truth is due to thought-ego, and this procedure destroys all innovation.

The answer to all challenges is the "awakening of consciousness", and we have to work on "objective observing and understanding" to regenerate the process of innovation.

Seeing is direct, hearing is direct, touching is direct, but thinking is indirect. Direct perception is true, the present, the fact, the non-duality, the what is. The truth about anything can be known by awareness. We have to learn to listen, to see with an open mind and without any resistance.

Ego: Albert Einstein stated, "The experiences of an individual appear to us arranged in a series of events; in this series the single events which we remember appear to be ordered according to the criterion of 'earlier' and 'later'. There exists, therefore, for the individual, an I-time, or subjective time. This is itself is not measureable".[1] The urge of I-consciousness, or psychological time or ego has not got any objective reality. These are just concepts, constructs of the mind's relative frame of reference. Psychological needs are human creations.

The ego-superimposes itself on thought structure, so thought-ego appear together and cannot be separated.

Dualism: Spencer Brown,, in a most spectacular passage, wrote, "In order to do so [to see or hear], evidently it must first cut self up into at least one state which sees, and at least one other state which is seen. In this severed and mutilated condition, whatever is seen is only partially itself". Sullivan observed, "we cannot observe the cause of nature without observing it", or Andrade, "Observation means interference with what we are observing .... Observation disturbs reality". This split of the mind to see and to be seen is called "dualism", it demands a rigorous, consistent, and persistent technology [not methodology] capable of pursuing dualism in its limits. D.S. Kothari wrote, "What is observed in different from what was before the act of observation". It is therefore next to impossible to uproot by intellection.

Psychological Imbalances: The mind constructs psychological responses, which are invariably false – say, fear, violence, terrorism, acquisition, and so on – and these are called psychological imbalances.

The innovated structures are man-made patterns and are false. So obviously man has started living in a mutilated or deformed world and the reality goes to abeyance. This has got to be corrected through acquiring higher awareness. At the moment of attention, there is no image, no thought-ego structure and no duality. The whole mind-brain-body system revels in Truth [Intelligence].

Reference

1. Albert Einstein, in Sampooran Singh, *The Holistic Education*, 1987, in L. Barnett, *The Universe and Dr. Enstein*, Victor Gollanez Ltd., London 1949, pp. 39-40.

14

Rationalism and the March of Reason

1. Rationalism and the March of Reason

Reason is a precious value thrown up by evolution and the source of much human progress in culture and civilisation. Swami Vivekananda wrote; "On reason we must have to lay our foundation; we must follow reason as far as it leads; and when reason fails, reason itself will show us the way to the highest plane".[1] The limitations of reason have been overcome by reason itself.

Bertrand Russell summarized St. Augustine's opinion in words, "Past and future can only be thought of as present: *'past'* must be identified with memory, and *'future'* with expectation, memory and expectation being both present facts". This view is supported by Alan W. Watts in words, "From memories you infer that there have been past events. But you are not aware of any past events. You know the past only in the present and as part of the present".[2] Schrodinger states, "There is only a *now* that includes memories and expectations".[3] This suggests man has trained his mind to graft the 'past' on the 'present', and he is aware of the 'distorted perception', which is *non-fact,* the *false,* the *past,* the *what should be?*

The world crisis demands radical revolution. Our problems demand it. This has resulted into ecological crisis. The Union of Concerned Scientists [over 1670 scientists including 104 Nobel Laureates] have issued *"World Scientists' Warning to Humanity"* on April 1993 stating, "A new ethic is required – a new attitude towards discharging our responsibility for caring for ourselves and for the earth ... The ethic must motivate a great movements, convincing reluctant leaders and reluctant governments and reluctant peoples themselves to effect the needed changes".[4]

This article explores rationalism and the march of reason. The main basis of the limitations of scientific reason in its sole dependence on the observed sense data of the external world and neglect of the observer or experience of the inner world. It also explores the meaning of life – the facts of psychological and trans psychological truth behind these facts, the absolute ground of our being, the ground of existence. It also discusses observation, insight, freedom, peace, and spirituality. The only concern of an enquirer is to understand the truth, in order to be able to discriminate the true from the false.

2. The Emergence of "I-Consciousness" or Ego

Man had an evolutionary origin. On the biological level, he had a separate body, which was relatively true. Man extended the idea of separateness and independent physical level to the psychological world. Man had a *self-conscious energy,* and this gave birth to ego. The self-conscious energy is used for thinking, feeling, reacting. Reaction is an emotional movement, reaction indicates a chemical disturbance in your system. This leads to neuro-chemical imbalance in your being, the perceptions and responses are distorted.[5] You lose peace and equipoise of mind.

J. Krishnamurti wrote, "The story of all mankind is me .... I am an individual, not the collective ... We are the history of mankind".[6] Hideki Yukawa, a Japanese Nobel Laureate in Physics, wrote, "Man has things in him that are unknown even to himself that he possessed before man became man".[7] There is a self-conscious energy moving through all of us – through the brain of each human being. The content of thought has been given to me, fed into the brain by the human race. The I-Consciousness, the ego has no factual reality at all. It has a conceptual reality. It has an ideational reality.

The "I" creates psychological needs, which are human creations. The "I" also creates competition. When ego enters the innovative process/product, it has a damaging effect. This reduces direct perception to a repetitive process and leads to distorted perception. This hinders the progress of innovation process/product.

Seeing is direct, hearing is direct, touching is direct, but thinking is indirect. The world is seen not as it is, but in various relationships to the "me" of memory. The past is the "me", so the *past* grafts itself on the *present,* and this results in *distorted present.* There is no "me" in the *present.* The scientific mode of mind is the non-dual and non-conceptual, so it is aware of the *present,* and is free from the "me". This gives us a technology to be free from ego – one has to be aware from the beginning to the end, not to become inattentive in the middle of it. This new quality of awareness is attention and in this attention, the "me" is absent. Observing and understanding is in the non-dual frame of the Mind Continuum, so understanding is not an intellectual affair.

3. A Critique on Mutation and Innovation

Mutation: Transformation is not in the future, can never be in the future. It can only be *now* from moment to moment. The *now,* the *present* is invariably in the non-dual and non-conceptual frame of reference of Mind Continuum. Only the truth can liberate the mind from its own ideation, to see the truth, the mind must realize the fact that so long as it is agitated, it can have no understanding. In stillness of the mind there is no projection of thought. Seeing the false as the false, and the true as the true is transformation, because when you see very clearly as truth, the truth liberates. When you see something as false that false thing drops away.[8]

Mutation is a sudden, abrupt, qualitative change in the totality of the mutant. It means radical qualitative change in the totality of our consciousness. Thought is invariably repetitive, transient, limited, dualistic, and past. When thought acts in this *past*, which is acting as memory, as experience, as information [knowledge], as opportunity. All will is desire based on this past and directed towards pleasure or the avoidance of pain. There is no new living at all. It means that psychologically the mind is sterile to bring any transformation. When the mind is quiet, silenced equipoised, the mind attains a higher quantum energy potential, and the new comes into being. All continuity is thought; where there is continuity, there is nothing new. Thought is not capable to bring any transformation in time.

Innovation: In the innovative process, there is no change in the successive or sequential consciousness, or in the time field. When the total *content* of successive consciousness goes to abeyance, there is instantaneous quantum jump from successive consciousness to Absolute Rest Frame, from time field to timeless realm, from dualistic frame of reference in the mind spectrum to non-dual and non-conceptual frame of reference in the Mind Continuum, from psychological mode to scientific mode; and this ends all activity of the innovation. To be precise, ***the innovation is in the time field, while the mutation is in the timeless realm; and mutation is invariably synergistic to innovation.***

Thought: We call the boundary, or the wall, between the symbolic-dualistic frame of mind spectrum and the non-dual and non-conceptual frame of reference of Mind Continuum as the "frequency barrier", or the "threshold frequency barrier", or the "event horizon". And events below the frequency barrier is within the speed of light and, any happening above the frequency barrier is beyond the speed of light. The innovation is bound by "speed of light", while the mutation is "*beyond the speed of light*". Thought can never penetrate the problem of human relationship. Thought is superficial and old and is the out come of the past. The past cannot enter into something that is totally new. Thought is the world, the symbol, the image. Thought is divisive, fragmentary and whatever it does will be separative and contradictory. It always sows the destruction, of war, of violence. To be precise, thought is sterile to bring any transformation in innovation. Further, mutation is in the timeless realm.

4. The Art of Observation

R.P. Kaushik wrote, "The first stage prior to recognition and naming is a state of non-verbal or though-free perception [*nirvikalpa jnana* or pure perception]. From this pure perception we proceed to thought, recognition and naming [*savikalpa jnana*]. When the observer identifies the observed, it classifies and qualifies it. However, in the psychological field, both these processes of identification and alienation are one and the same. Hence, there is no observation but only distortion.

When you observe without identification, recognition, naming and non-verbally, there is "direct perception". Observation, understanding, learning lead to pure perception. Perception has nothing to do with knowledge [information]. ***The perceptions are born of non-duality and responses are born of spontaneity.*** It awakens the awareness, the Intelligence, which arise out of observation, self-knowing and compassion.

Learning: Learning is possible when one is attentive. Attentivity implies awareness of the fact, the whole body must be aware of the fact, the wholeness. Attentivity is possible when one is not tuned to the chatter of the brain, but tuned to the set of frequencies of the fact. Learning is possible by encoding the information as wholes, it is possible when there is perception of the fact.

Understanding: It is understanding that flows to learning. Understanding and learning blossom to spontaneity and right action. It flowers to ending the movement of pain, strife, conflict, and so on. The intuition or insight can decondition the mind-brain-body system. Insight is perception of the wholeness and from that completeness there can be logical, sane, rational action. Only then can you have a complete relationship with the mind. The consummation of human life is a pilgrimage to the sacred temple of understanding, learning and right action.

5. Abstraction and Intuition

Maurice H. Wilkins wrote, "Science makes progress largely by a combination of two aspects: open-mindedness and the fallowing of abstract concepts .... Science always puts the most emphasis on this abstraction and narrowing down, on analytical procedure and on the role of the intellect as distinct from intuition and feelings .... The logical and rational is very important element in science. But emphasis on that tends to make people not notice the essential role of intuition".[9] Hideki Yukawa wrote, "The important point would appear to be the balance between intuition and abstraction"[10] He added, "It seems as if present-day physicists have lost the gift of foresight inherited from their forerunner".[10]

Intuition: Throughout the ages man has recognised two states of mind – self conscious mind and intuitive mind. If a man meets a sudden and unexpected danger – say a serpent in the middle of the road or the house on fire – he jumps to a place of safety without thinking and this is called intuition. The intellect or reason goes to abeyance in a non-dualistic mind and the inner perception takes over momentarily. This is nature's grace.

Intuition or insight is the timeless perception. It is insight alone that can recondition the brain. Insight is complete and from that completeness there can be logical, sane and rational action. Only then there can be flowering of the brain. Pure clear insight is perception without any doubt. ***Understanding, learning and action are a cyclic process, which results in a spiral going upwards in the vertical dimension; it is a growth in the vertical dimension***. The higher the understanding, the higher the expression of the Consciousness in the human brain-body system.[11]

Creative Perception: Galileo Galilei wrote that, "The mere acceptance of stimuli which come from the environment and which results in the accumulation of memory in the brain is insufficient for our purpose. He taught us by his own achievement that experiments purposely designed are of vital importance in uncovering the truth hidden in nature. He asked questions addressed to nature and succeeded in receiving answers directly from nature".[12]

Herman Weyl wrote, "Understanding from within nature is open to us .... of myself, of my own acts of perception, thought, volition, feeling and doing. I have a direct knowledge entirely different from the theoretical knowledge that represents the 'parallel' cerebral processes in symbols. This inner awareness of myself is the basis of understanding of my fellow men".[13] Ilya Prigogine and Isabelle Stengers completely agree with the above.

Voltaire wrote the following about destiny.

.... everything is governed by immutable laws .... everything is prearranged .... everything is a necessary effect .... both are toys in the hands of destiny.[14]

Alexander Koyre wrote, "A world, in which there is place for everything, there is no place for man .... This is the tragedy of modern mind which solved the riddle of the universe, but only to replace it by another riddle; the riddle of itself".[15]

6. Freedom

It is pure observation without direction, without fear of punishment and reward. Freedom is without motive; freedom is not at the end of evolution of man but lies in his first step of his existence. In observation one begins to discover the lack of freedom. Freedom is found in the choice of awareness, in the attentiveness of our daily existence and activity. Freedom comes, when even momentarily, thought or pseudo-subject goes to abeyance and attentiveness takes over as the true-subject. Pure observation bestows insight, which is without any shadow of past or time. The timeless insight brings about a deep radical mutation in the conditioned psyche or mind-brain-body system.

Understanding the nature of bondage is the beginning of freedom. Understanding the content of bondage gets converted into factual urge for freedom. Freedom is spontaneous equibalance or equipoise of mind. The human being seems to be born with the faculty of self-awareness and understanding, bondage and freedom, dualism and non-dualism. Thought-time bondage of dualism are in the field of the known; and understanding, freedom and non-dualism are in the realm of unknown.

The aspiration of unconditional freedom in every individual is born when one transcends the dualism of mind-brain-body system which flows into dropping the unreal and the process of becoming comes to an end. This results in a harmonious, non-violent and peaceful life.[16]

Peace: It is a total way of living. As long as there is inherent violence in the human psyche, there can never be sustainable peace in the world. Peace requires the mutation of individual psyche. **It is only holistic education that helps**

man to grow from knowledge to understanding, from confrontation to cooperation, from rationality to intuition, from intellect to intelligence.

We are on the crossroads of destiny; we must take the road from information gathering and experience to understanding and learning, from violence to nonviolence, to motivation of sharing rather than to dominating, complementary and not competition, harmony and not discord; and in the final analysis education alone that can bring this about.

Peace is a state, when thought-noise ends, a frame of reference of the Mind Continuum. A sustainable peace is defined as when the mind attains the capacity to remain in the state of attention or a state of awareness. Peace requires free human beings. The search for peace would be a search of new dimension of consciousness.

We want a psychological revolution, in which the whole human race has to participate and every man is an equal partner. This collective psychological experiment will awaken the integral and all-inclusive dimension of Consciousness and that can guarantee human survival along with all fauna and flora.

7. A Vision For the Twenty First Century

President A.P.J. Abdul Kalam in a seminar at IIT, Delhi on March 31, 2005 suggested intake of about 300 M.Sc. and 100 Ph.D. scientists with proper emoluments and assured career growth in organisations. R. Chidambaram visited Guru Nanak Dev University on May 17, 2006 said, "India needs to build scientific temperament for creating science talent pool among the young scientists for a strong nation". V. Krishanamurthy, in *"THE HINDU Nov. 17, 2006"* also stressed on innovation in manufacturing. ***Nearly six decades after independence, neither we have a school to explore human mind, nor inward transformation or inward flowering***

and character building through educational structure – "to allow teaching within a spiritual and holistic content".

As India moves into the sixtieth year of Independence, it is finding itself adrift, aimless and almost rudderless. The Editor [H.K. Dua] of the Tribune of August 15, 2006 wrote in historic words:

The tragedy is that those who are clear about their vision – they are only a few in number – are finding themselves unable to take others along; and those who do not have the vision but only personal interest to guide them.

P Krishna, on April 3, 1993 said, "We are going towards self-annihilation and not much time is left now".

8. Concluding Comment

Albert Einstein, in prophetic words, wrote that we need "Objective observing and understanding", and this leads to "peace and serenity". When nature is questioned, it provides answer as direct perception; and this is called "discovery". Discovery is in the *timeless* realm [non-dual and non-conceptual frame of reference of Mind Continuum] and innovation is in the field of time [dualistic frame of reference], so discovery flows to innovation.

Creative leadership emerges when man discovers new things hidden in nature. Man has to begin with intuition [insight], only then he can be creative. Intuition and discovery are synonymous. Intuition is fundamental and reason [innovation] is a derivative.

The subjective research is a dialogue with nature. There cannot be any innovation without first discovery. "We as a nation will have to innovate to survive"; then the highest priority is to *understand* ourselves and live that understanding. There is an urgency to restructure the dimension of science –

[Science-Education-Ethics-Health] – in mankind *here* and *now*. We cannot afford to waste even a single nanosecond.

References

1. Swami Vivekananda, *Complete Works, Vol. 1 [eleventh edition]*, p. 185; in Swami Ranganathananda, *Science and Religion*, Advaita Ashrama, 2005, p. 109.
2. Alan W. Watts, *The Wisdom of Insecurity*, New York: Vintage, 1968, p. 82; in Ken Wilber, *The spectrum of Consciousness*, USA: The Theosophical Publishing House, 1982 [Quest Book], p. 98.
3. Erwin Schrodinger, *What is Life? & Mind and Matter*, London: Cambridge University Press, 1967, p. 145.
4. USA: The Union of Concerned Scientists, *World Scientists' Warning to Humanity*, April 1993.
5. Vimala Thakar, *Passion for Life*, Bookfund: Holland, 1989, p. 95.
6. J. Krishnamurti, K.F.I., 1982, Bulletin I, p. 40; K.F.I. 1983, No. 3, pp. 27, 28.
7. Hideki Yukawa, *Creativity and Intuition*, Tokyo: Kodansha International Ltd., 1973, pp. 205-206.
8. J. Krishnamurti, *The First and Last Freedom*, London: Victor Gollancz Ltd., 1977, pp. 278-279, 286.
9. Maurice H. Wilkins, *Synthesis of Science Religion*, Mumbai: The Bhaktivedanta Institute, June 1988, pp. 30, 31.
10. Hideki Yukawa, *op. cit.*, pp. 101-102, 105.
11. Sampooran Singh, *A Review on Symbiosis of Science and Spirituality*, Dagshai: *Atam Science Trust*, October 7, 2002.
12. Galileo Galilei, in Hideki Yukawa, *op. cit.*, p. 111.
13. Herman Weyl, *Philosophy of Mathematics and Natural Science*, Princeton University, Press, 1949, in Ilya Prigogine, and Isabelle Strengers, *Order Out of Chaos*, Bantam Books, 1984, p. 311.
14. Voltaire, *Dictinnaire Philosophique*, Paris: Garnier, 1954, in Ilya Prigogine, *ibid.*, p. 257.
15. Alexander Koyre, *Newtonian Studies*, Chicago: University of Chicago, 1968, pp 23-24, in Iya Prigogine, *op. cit.*, p. 36.

16. Sampooran Singh, *Education for a Global Society,* Dagshai *Atam Science Trust,* October 7, 2001.

15

The Act of Living

1. Man Started Living in Distorted Present

Man had an evolutionary origin. He started observing, comparing and seeking to classify his observations at least for more than a million years ago. This gave birth to crystallise behaviour patterns or "cultural belief structures" or memory as frequency and amplitude information. He also trained his mind to a coherent superpositions of different webs [past] on the present resulting in a perception of the *"distorted present"*, where "past and future can only be thought of as *present*: *past* must be identified with memory and *future* with expectation, memory and expectation being both distorted present facts".

Spencer Brown, in a most spectacular passage wrote, :"In order to see, evidently it must cut itself upto into at least one state which sees, and at least one other state which is seen. In this severed and mutilated condition, whatever it sees is only partially itself". Arthur Eddington put it succinctly, "Nature, thus, provides that knowledge of one-half of the world will ensure ignorance of the other half." This "severed and mutilated condition" is open to environment and closed to the Absolute, the Nature. There is exchange of energy between the subject and the object and both are distorted. We cannot observe the course of nature without observing it. To live in a distorted world, we call it as the act of living. This constructs violence, strife, our psychological imbalances, and we are caught in nervous tension and chemical pressure.

The crisis is in the psyche, in the human consciousness and human consciousness has accepted the existence of "I-ness", the "me", the "ego", the "self". ***It has no factual reality at all, it has a conceptual reality, it has an ideational reality,*** and it has a relevance in our lives. In other words, the concepts [recalled memory, past] are grafted on the perception [present]; man has started living in distorted perception, because there is nothing like the "I", the ego, the factual reality.

The energy, which sustains the "I" process, is your identification, attachment and addiction to it; it is an involuntary process. When you manufacture, market, sales a product, you enter into competition by identifying and get attached to the product. This brings in "I" consciousness in the innovation process, and this retards the progress of innovation.

This article explores the act of living in distorted present. It also explores the ending of dualism, or the ending of ego, and the Unconditioned Energy. It is aware of the process of innovation. It discusses education, subjective research and ending of ego. It is the higher quantum energy potential of consciousness that has the healing effect.

2. Innovation

What we call mind [with a lower case of 'm'] is a conditioned energy moving through the body, monitored by brain, operating through the senses. The mind is a conditioned conscious energy. It is conditioned by the total human race through untold centuries. It generally deals with thought-structures which are invariably entangled with volitions, emotions, sentiments, desires, fears, violence, and so on. Conditioning implies the lowest hierarchical expression of consciousness, or the lowest quantum energy potential of consciousness of mind-brain-body system.

The conditioned psyche, in order to see itself, evidently it first cuts itself up into at least one state which sees, and at least one other state which is seen – or in other words, it divides itself as subject and the object, thus creating dualism. The error of dualism forms the root of intellection and is therefore next to impossible to uproot by intellection. The distorted subject and the object are at the same quantum energy potential. To detect this demands a rigorous, consistent, and persistent technology [not methodology] capable of pursuing dualism to its limits. The content of thought has been given to me, fed into the brain by the human race. The cerebral movement taking place in the mind-brain-body system having the content conditioned by humanity is not individual, it is something collective. It is shared by the whole mankind. So, the "I" consciousness goes on creating needs and wants, pleasure and pain, likes and dislikes.

Innovation is a new idea – emerging from invention or discovery – and this is converted to a commercially useful process/product. It is a materialistic pursuit and is governed by thought process. Innovation is defined as a movement of "quantum energy potential of consciousness" from an unfinished process/product to a commercially useful product – as the final product is at a higher order than the unfinished product. The mind divides itself into the subject and the object where both are at the same quantum energy potential, or in other words, there is no transfer of energy between the subject and the object – this is called dualism. Acceptance of the authority of the thought process, the experiencing process, the acquisitive process, through the brain or through the neuro-chemical system keeps you tethered to the centre, the monitor that you call the ego or the "I" consciousness, very quietly, very stealthily in a subtle way, without expressing its sting of pride and vanity.

If observation is again continued, the mind moves from knowledge to acquaintance, to a personal encounter with the fact through observation. The ego seems to play a very important role in the pursuit of innovation. Man, unknowingly, tried for many thousand years, but have practically failed.

Recently Henry Chesbrough, in his book *"Open Innovation"* [2003] talked about open innovation as a paradigm shift from the previously practiced closed innovation wherein companies used to think themselves as self-sufficient in their innovative competence [reviewed by Sandhya, U., in the Hindu, Delhi, November 8, 2006]. The innovation without deep Understanding of the creative energies [discoveries/inventions] and the role of *"ego"* create an illusion.

3. A Critique on Education

The educational-structure in India and in the rest of the world are rote methods of learning and intellect [reason] is supreme. One of the unsolved problems and challenges to educationist all over the world in the problem of freedom and order. The five devils are population increase, fast depletion of earth resources, ecological problems such as pollution of air, water and earth, economically and unbalanced world and arms race or expenditure an armaments. All students are thrown into the world of strife, confusion, violence, war and hatred. The recent problems are terrorism, drug trafficking, and religious fundamentalism. Our values are money, position, prestige, power. Man is growing with pain and misery.

The eminent philosophers and educationists – Swami Vivekananda, Sri Aurobindo, Radhakrishnan, Zakir Hussain, Mahatma Gandhi – opined that *"no education worthy of the name can be conceived as something external, having no relationship with man's innermost being and becoming"* and task of education is inward transformation or inward flowering, or character building. The foundation of education is not knowledge [outer] but intuition [inner]. Education is to cultivate and nourish the self-integrative potential [non-dualistic frame] of Mind Continuum; and to jettison the self-assertive tendency [dualistic frame] of mind spectrum.

Education implies educating the body, educating the psychological mode and trans-psychological mode, intuition, invention and discovery. We have to awaken the mind. We have to learn to live with tremendous awareness. We have to educate ourselves in the dimension of Silence and energy of Intelligence. Education is of prime significance for creation of a new culture, a new human being. The scientific and Medical Network in UK and in Europe has made fantastic progress, but we in India are lagging behind.

4. Subjective Research

Thought is intellect which is a cerebral energy, it is created by the past inheritance. Impulse is a biological energy. The non-dual and non-conceptual frame of reference is synergistic to thought. Intelligence is an energy born of the emptiness of *silence,* the space of *silence.* Intelligence is perceptive sensitivity. Intelligence is the foundation of subjective research.

Physics tells us that: (a) the world extended in space and time is but our representation; (b) the objective scientific view of natural processes formed in this way lack all sensual qualities; (c) the subject of cognizance goes to abeyance [Principle of Objectivity] in objective research, so it moves from *fact* to *fact*; (d) many eminent scientists – Karl Pribram, Roger Sperry, John Eccles, Wilder Penfield – advocated that we consider the objective and the subjective as a single dynamic flow; and (e) if we follow a rigorous, consistent and persistent technology, then we have to restructure science – [Science-Education-Ethics-Health] – which deals with the wholeness of life and bestows on us the 'subjective truth', or 'pure perception' or 'timeless perception', or 'discovery', which flowers to order, harmony, nonviolence and peace.

In subjective research, the violence, the conflict and all psychological imbalances go to abeyance, so man revels in peace and serenity. He assumes the role of a co-worker with the Divine.

5. The Ending of Ego

a) The ancient seers of India, about 10,000-30,000 years ago, perceived a state of One Reality – the Unmanifest-Manifested. The dimensions of the Unmanifest [Passive Reality] are Homogeneous, Undifferentiated and Non-dimensional; while of Manifest [Active Reality] are Heterogeneous, Differentiated and Multi-dimensional. The Active Reality is a powerful, highly directional, monochromatic [same frequency] and coherent [the wave is in one phase] beam of energy [Consciousness], or a state of the highest frequency [higher frequency means higher energy] or highest quantum energy potential of Consciousness. The intense energy of insight expresses itself as direct perception of the fact, the Reality. The Active Reality is often called the Subjective Research.

When we educate ourselves in the grace of non-motion, non-action, in the beauty of relaxation, we expose the ego to the intense energy of insight, and ego goes to abeyance. It is a non-cerebral, non-psychological exploration of the conditioned psyche, the ego. One will have to perceive energy moment with tremendous alertness and sensitivity putting your whole being into that action, into that moment, in every relationship. This alertness enable the mind to go to a quantum jump from dualistic state to non-dual and non-conceptual state which is called the scientific temper. The subject of cognizance goes to abeyance, and one lives from *fact* to *fact*. The non-dual state is free from any ego, and it leads to intuition [insight]. The intuition is also called "pure perception" or "direct perception".

It is the consciousness, the awareness, the alertness that does not require duality – the division of life into the subject and the object, the observer and the observed, the doer and the action. Silence that blossoms to non-duality, relaxation, energies the being. And the being is free from all violence, fear, psychological imbalances, and revels in freedom, peace and bliss. The consummation of human growth is freedom from all senses of belonging materially, physically, verbally and psychologically.

The innovation process [new manufacture] has to be carried over by the scientific mode of mind; but the business activity [as competition, marketing, sales] has to be entrusted to another group of persons, trained as technocrats.

(b) **Ending of Psychological Imbalance**: All human problem – violence, conflict, terrorism, nervous tension, chemical pressure, etc. – are construct of the conditioned mind, so these cannot be resolved by knowledge and thought at the cerebral level. All cerebral activity is at the dualistic level, so the relative frame of reference is sterile in solving human problems. The "me", the "ego" is a part of thought, so the ego destroys the innovation process. To end all misery at the cerebral level, we have to conduct an enquiry in utter freedom in a non-authoritarian way. The enquiry must raise the potential of awareness and lead it to scientific temper. As the enquiry proceeds it makes minor quantum jumps in the level of awareness and this is shown in Fig. 2. The dotted lines show the quantum jump to an higher level of awareness. When you conduct an enquiry with each step there is a movement away from the false in our daily life, in our daily relationships; in other words, it means that is higher and higher level of awareness. When all matter content of innovation goes to abeyance, it makes quantum jump from dualistic frame to non-dualistic frame of reference of Mind Continuum, which blossoms to intuition and it provide an answer to the challenges, the innovation.

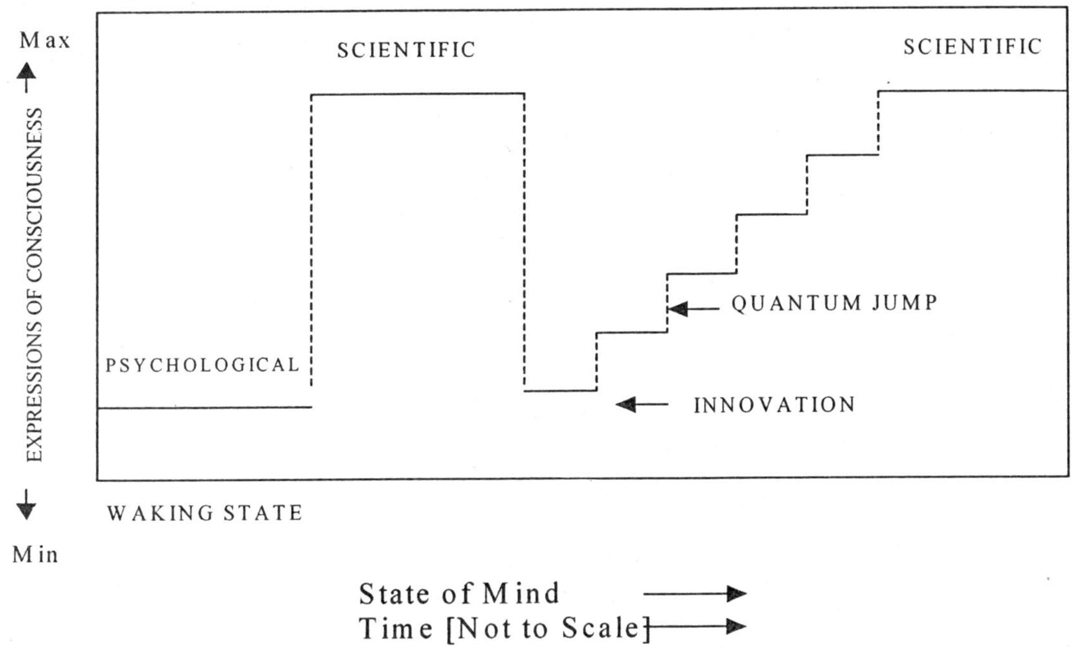

Fig. 2: A series of minor quantum jumps from state of innovation to scientific temper. This results in ending of ego at the level of scientific temper. The dotted lines show the quantum jump, which happens from psychological mind to scientific mind.

(c) **Action Plan** : a) There is nothing mine, it is an organised stream of knowledge, evaluations, criteria, values-structures. All of them are moving through the bed of my body. There is no reaction-free observation.

(b) The thought and the "me" cannot be separated, both are coupled. The movement of thought and the "me" are not two separate things, but are one and the same.

(c) When I observe my thoughts, I am observing the organised collective thoughts of the whole human race and I am a part of that. The "me" is a part of that; the "me" cannot be separated from that movement.

(d) When you conduct an enquiry, with each step there is a movement away from the false in our daily life, in our daily relationships. The false drops itself. This is called minor quantum jump. When I learn observation and the reaction-free attentiveness comes to be there as a steady flame in my consciousness, and many secondary and false things drop away. The distinction of subject and object is utterly obliterated, when the enquiry and the enquired are fused into one perfect identity. There is a quantum jump from dualistic frame to non-dual frame, from psychological mode of mind to scientific temper. This bestows *objective truth* and the innovation blossoms.

(e) Observation is possible in the reaction-free attentiveness. Observation and understanding are in the non-dual realm.[1] It is awareness that rejuvenates the innovation.

6. Concluding Comment

The symbolic-dualistic frame of reference of mind is local, bounded and limited. Thought is always limited. The thought and dualism construct "I", ego. The innovation is a pursuit of materialism, and ego acts as hindrance in its pursuit. The ego has no factual reality at all. It has a conceptual reality. It has an ideational reality.

The foundation of education is not information [knowledge] but intuition [inner]. It is self-knowledge which leads a dualistic mind to make a quantum jump to non-dualistic frame, The non-dual and non-conceptual frame of reference is called the scientific temper.

It is only the right education that can bring the revolution between man and man, between human beings and between man and nature. It is only the action of will and assertiveness that needs to be changed at all levels. A new human being must be free from the weight of time, free from the burden of the past, and free from all the actions of will. This is the new human being. This only is the social, psychological, and even the political revolution.

A depth probe suggests that the roots of our cultural crisis lies in an imbalance between – one, the intellectual process, the observer operating on the thing observed which is analysis, which is correction, the altering and adding; and the other is direct perception, the *fact,* the intuition.

The objective science moves from *fact* to *fact*. The Universal Consciousness express itself in a non-dual frame as intuition [insight], spontaneity, love and compassion. The objective mind has a capacity to make another quantum jump from outside field to *inner realm*, the Universal Consciousness sees Itself in the inner realm, and gives birth to subjective science.

When the innovation is observed by the intense energy of insight, the ego goes to abeyance and the whole system reverts to scientific temper. The survival of human race is embedded in our faculty of Intution, Invention and Discovery.

The energy of intelligence is called perceptive sensitivity. The intelligence is born from the silence. It is direct perception or pure perception. The intelligence is free from all violence, fear, conflict, psychological imbalance; it revels in Love Truth and Peace. Discovery is fundamental and innovation is a derivative.

Reference

1. Vimala Thakar, *The Benediction of Being Alive,* Bookfund Vimala Thakar, The Netherlands, April 1984, Nr. 15, pp. 8-41.

16

Consciousness, Creativity and Innovation

1. What is Consciousness?

The ancient seers of India, about 10,000 – 30,000 years ago, perceived a state of true illumination. There is One Reality, the Unmanifest-Manifested. The Active Reality is a powerful, highly directional, monochromatic [same frequency] and coherent [the wave is in one phase] beam of energy; or a state of highest frequency [higher frequency means higher energy], or high quantum energy potential. The intense energy of insight expresses itself as direct perception of the fact, the Reality.

The Universal Consciousness or Reality [highest quantum energy potential] heals, because it has the dimensions of creativity and spontaneity, nonviolence, beauty and harmony, love and compassion. Man has to choose between healing and annihilation between the *Buddha* [Wisdom] and the *bomb* [annihilation]. Obviously the decision is to work honestly, sincerely and truthfully for mutation of one's own psyche.

Consciousness performs a dual role in the universe. It is the *being* and the *knowing* of experience and information [knowledge]. In the world of the quantum, it is both the awareness and the creation of experience. With the stroke of the twentieth-century quantum energy eraser, the dividing line between ontology [theory of being] and epistemology [theory of knowing] is rubbed out. The fusion of ontology, epistemology and psychology is beginning to emerge.

Modern science has shown that all violence, conflict, terrorism, psychological imbalances are constructs of the conditioned mind-brain-body system. And obviously we have to explore the conditioned psyche to end these challenges.

Prime Minister Dr. Manmohan Singh said on August 2, 2005 that, "On an average, more than 3,50,000 engineers and 5,000 Ph. D. scholars graduate from our universities and colleges every year". We can enhance our R & D effort many-folds if we understand the "Creativity and Innovation.

The world is a construct of our sensations perceptions, memories. Every cerebral process is an expression of consciousness. All creativity is an expression of consciousness at different quantitum energy potentials. Consciousness is never expressed in the plural. This article explores, *"What is Consciousness"*, the manifestations of creativity, the expression of innovation, ending of ego, ending of violence, new relationships between creativity and innovation and intelligence.

2. The Manifestation of Creativity

a) Hideki Yukawa, Japanese Nobel laureate in physics, wrote, "Creativity implies discovering something that nobody has hitherto known about or inventing something new .... Invention and discovery always have something of the nature of the unpredictable". He added, "Prior to creativity there has to be a struggle with one's own self .... There has always existed the possibility of such creativity appearing, that something hidden, something latent, should appear, should become

manifest".[1] In scientific terminology, we say that *"intuition"* **makes** a quantum jump to higher level of awareness as *"invention*; and invention has the potential of another quantum jump to still higher level of awareness as *"discovery"*. All *intuition, invention, discoveries* were expressions of the Consciousness, the Unmanifest. Yukawa stated that in nature all intuition and abstraction [scientific] were in complete harmony and balance with each other.

All great discoveries have arisen from the realm of the timelessness, so scientific enquiry is not a cerebral activity and gathering information, but a quest of the Timelessness, the Divinity, the Life Field. A selfless, unfaltering devotion to truth and use of knowledge [perception] for man's uplift – higher hierarchical expression of consciousness – are the only vocation of a scientist.

Erwin Schrodinger, a Nobel laureate in Physics, 1931, stated, "I consider science an integrating part of our endeavour to answer the one great philosophical question which embraces all others, the one that Plotinus expressed by his belief: *"Who are we really? Where have I come from and where am I going"*.[2] And more than that I consider this not one of the tasks, but the task of science, the only one that really counts".[2] Maurice H. Wilkins, Nobel Prize in Medicine, 1962, wrote, "the open-minded enquiry of the scientist is not something peculiar to science itself but is a characteristic of the good way of living for human beings in general".[3]

(b) ***Geniuses Appear in Batches:*** Yukawa wrote, "It is usual, it seems for geniuses to appear in batches; on the other hand, there are periods during which the appearance of geniuses is very rare". It appears that nature stirs up the ocean of Universal Consciousness in a certain time-period, which appears that geniuses come in batches, which in other time-period, nature is passive and the appearance of geniuses is very rare. Nature thus controls, the Wisdom or Intelligence in the universe.

Voltaire wrote, ".... everything is governed by immutable laws .... everything is prearranged .... everything is a necessary effect .... we both are equally foolish, both toys in the hands of destiny".[4]

(c) ***The Art of Creativity:*** We discussed life histories of some eminent scientists – Galileo Galilei [1564-1642]; Gregor Johnn Mendel [1822-1884]; Albert Einstein [1879-1955]; Srinivasa Ramanujan [1887-1920]; Shankaracharya [686-718][5]. This shows that truth self-manifest in a pure mind. All perception [knowledge] is structured in the consciousness, it is embedded *"within"* of man. Man is a symbol of perfection, an emblem of the Divine, but he has to reveal this perception by shedding his ignorance. Man is a conscious being who have an hidden capacity to reveal the Divinity, the Life Field. The entire human race is a creative unity of human individuals.

Karl Popper expressed the quintessence of the method of science as: [$P_1 \rightarrow TT \rightarrow EE \rightarrow P_2$]. Here P_1 stands for the problems which serve to test the theory; the testing of the theory is symbolized by TT. The testing leads to elimination of errors [EE] in the theory. The improved theory suggests new problems P_2 and the process repeats itself.

Albert Einstein recognised the movement of the frame of reference of the Mind Continuum. In almost the last interview that he gave to the international magazine Life said:

"The mind can proceed only so far upon what it knows and can prove. There comes a point where the mind takes a leap – call it intuition or what you will – and come out upon a higher plane of knowledge, but can never prove how it got there. All great discoveries have involved such a help".

(d) ***Intuitive Perception:*** When the non-dual and non-conceptual frame of reference [scientific temper] is established in the Mind Continuum, it is stated to revel in intuition. Albert

Einstein says that all knowledge is intuitive in words, "Physics is essentially an intuitive and concrete science. Mathematics is only a means for expressing the laws that govern phenomena".

Einstein recognizes that our scientific knowledge is primitive as compared to the limitless and unfathomable knowledge of Reality or Universal Consciousness or God. He echoed these feelings in the words:

"It is enough for me to contemplate the mystery of conscious life perpetrating itself through all eternity, to reflect upon the marvelous structure of the universe which we can dimly perceive, and to try humbly to comprehend even an infinitesimal part of the intelligence manifested in nature".

"One thing I have learned in a long life: that all our science, measured against reality, is primitive and childlike – and yet it is the most precious thing we have".

"We know nothing about it (the nature of God, the universe and man) at all. Our knowledge is but the knowledge of school children .... Possibly, we shall know a little more than we do now. But the real nature of things – that we shall never know, never".

(e) ***Creativity:*** When the mind is in the realm of *intuition, invention, discovery,* the Mind is revelling in the non-dual and non-conceptual awareness; the Mind Continuum has nothing to do with the "me", the "thought-ego", with the physical and psychological inheritance. It is an unconditional energy born of the emptiness of *silence*. It is an energy born of the total unconditional relaxation. It is not a cerebral movement any more.

The creativity is expressed as *intuition, invention and discovery* – it is also called as *Intelligence or Wisdom.* When the creativity is awakened, the perception is born of the non-duality of the wholeness of your being. You are totally relaxed, out of that wholeness, the creativity, which is non-dual is born. It is not a cerebral movement any more, it has nothing to do with the "me", the "I-consciousness, with the physical or psychological inheritance. It is an energy born of the total unconditioned relaxation that takes place in *silence*. It has no past and no future.

(f) ***Innovation:*** Man is aware of his hidden faculties – as *intuition, invention* and *discovery* – for uncountable aeons. Man is basically an enquirer of "truth and meaning of the wholeness of Life". He has recognised that no human problem, no problem of human relationship can ever be resolved through knowledge, through thought, on the cerebral level. Because every activity at the cerebral level divides the mind into the subject and the object, and the whole system is caught in the grip of dualism resulting into nervous tension and chemical pressure.

The intuition, the invention, the discovery are in the non-dual and non-conceptual frame of reference of the Mind Continuum; which we call Mind [with upper case 'M']. In this realm, the frequency [higher frequency means higher energy] is more than the velocity of light, so everything is faster than the speed of light.

Innovation refers to a new idea in the dualistic realm. It is basically converting direct perception in the scientific temper, and then use it to a commercially useful process/product. The mind falls through *"the event horizon"* and enters the materialistic field, the dualistic field. The dualistic field is governed by "ego", the "I", the "me". The ego and psychological imbalances damage the innovation process. What we called mind [with a lower case 'm'] is a conditioned energy, moving through the body, monitored by brain, operating through the senses. It is a movement of the structured frame work of patterns and the dynamics of behaviour patterns are shared by all human beings. The time-space matrices represent the relative field of consciousness. It constructs pain, misery and travails. It leads man to self-annihilation.

J. Krishnamurti wrote, "So long as the experienced verbalizes the feeling, the experience, he separates himself from it and acts upon it, such action is an artificial, illusory action".[6] Innovation is not a functional activity; it is a business activity and you need every component of business – manufacturing, marketing, sales, profit. If the experiencer thinks of business, it takes the role of psychological thinking, and the subject of cognizance splits itself into the subject and the object, which destroys the innovation.

Knowledge is an intellectual activity; it cannot lead to direct perception. The innovation is a materialistic activity, dualistic activity. The innovation gives us "thought coupled with ego". So thought-ego destroys the innovation.

Thinking is always dualistic; it separates the thinker and thought. Both are at the same quantum energy potential of consciousness. Thought is always self-protecting, self-perpetuating, conditioned. Thinking has not solved our problems. Intellect cannot resolve the problems of innovation.

To sum up, the *intuition, invention and discovery* are Nature's response to the human mind. The innovation is a construct of the human mind. The *thought-time-ego* is a human construct, and there is no growth, no freedom and no discovery. It is the ego coupled with thought, including psychological imbalances, nervous tensions and chemical pressures, which are responsible for the failures of innovation. If man enquires into – objective observation and understanding – the limitations of his conditioned psyche, the Nature leads the conditioned psyche by bestowing higher awareness. All healing is by Nature – through awareness - and we are only toys where some one else [Nature] pulls the strings.

3. Ending of Violence

We have built a society, which is violent, and we, as human beings, are violent. All psychological imbalances – violence, fear, greed, anger and so on – are constructs of the waking state of mind, or relative frame of reference of mind in the mind spectrum. The socio-economic-political systems, the institutionalised religions, the reforms in the external field have so far failed to transform man.

Violence is an indication of psychological immaturity. When the scientific mode of mind sees the cause of psychological immaturity, all violence goes to abeyance and nonviolence self-expresses itself. We conclude that violence is a construct by symbolic-dualistic frame of reference, and nonviolence is a percept in the non-dual frame of reference of the mind spectrum; where the non-dual frame is synergistic or at higher quantum energy potential to symbolic-dualistic frame. This lays foundation of "humanist technology", or "science of life", or "subjective research".

Violence implies to be free of the inner aggression, hatred and anger in the consciousness? And this has a cause, all disorder has a cause, and if that cause is seen, understood and eliminated it is possible to end the disorder. And when that disorder is ended there is automatically order. You cannot impose the order on top of disorder. You cannot impose nonviolence on a consciousness, which is violent because that violence has deep-rooted causes in the psyche. Unless these causes are eliminated, there is no ending of violence. It is only where violence ends there can be nonviolence. The violence can be ended when we explore "objective observing and understanding".

Peace is a total way of living. As long as there is inherent violence in the human psyche, there can never be sustainable peace in the world. Peace requires the mutation of individual psyche. It is only holistic education that helps man to grow from knowledge to understanding, from confrontation to cooperation, from rationality to intuition, from intellect to intelligence.

We want a psychological revolution, in which the whole human race has to participate and every man is an equal partner. This collective psychological experiment will awaken the integral and all-inclusive dimension of Consciousness and that can guarantee human survival alongwith all fauna and flora.

4. New Relationships Between Creativity and Innovation

Creativity and innovation are essential to business success as we are set in the 21st century. Paul E. Plsek advocated "directed Creativity, which he defined as, "The deliberate mental action needed to produce novel ideas in targeted areas" [7]. But we intuit that creativity [awareness] cannot be *"directed"* as creativity is embedded in the non-dualistic frame of reference of Mind Continuum.

Life is relationship and relationship is action; it is expressed through contact with things, with people and the ideas. It is the lack of understanding of relationship that causes conflict. It is an interconnected challenge and response. Relationship means communion without fear, freedom to understand each other, to communicate directly. We need new relationship between man and nature, man and man, and man and environments. Relationship is really a process of self-revelation, which is a process of self-knowledge.

Albert Einstein stated, "Objective observing and understanding". We have to observe the whole process of innovation without identification, without motive, without acquisition – it is direct perception. It is the direct perception which can lead the innovation state to scientific temper. It is the higher hierarchical level of awareness that can result in radical transformation [mutation] of innovation. Such a scientific temper will blossom to intuition, and to give solution of radical transformation of innovation.

5. Intelligence

That inner space, free from the contamination of thought and its movement, is called silence. Silence is that inner space of the wholeness of your being. Intelligence is an unconditioned energy born of the emptiness of silence. It is not a cerebral movement any more, it has nothing to do with the "me", the I-consciousness, with the physical or psychological inheritance. It is an energy born of the total unconditional relaxation that takes place in silence. So the content of intelligence is the emptiness of silence. Intelligence has no past and future.

6. Concluding Comment

(a) No human problem, no problem of human relationship, can ever be resolved through knowledge [information], through thought, on the cerebral level. Because all cerebral activity is at the dualistic level, or at the relative frame of reference. We have to conduct an enquiry in utter freedom in a non-authoritarian way.

(b) The knowledge and the experience gathered by the conditioned psyche is holding the whole world at ransom, it is holding the whole world in bondage. Knowledge does not result in direct perception. India has lived all these centuries on ideas. She has to move from a life based on ideas, which are non-facts, to seeing of fact only. The scientific temper moves from *fact* to *fact.*

(c) When I observe my thoughts, I am observing organised collective thoughts of the whole human race and I am a part of that. The "I", the ego cannot be separated from that movement of thought.

(d) When you conduct an enquiry, with each step there is a movement away from the false in our daily life, in our daily relationships. When I learn observation, the reaction-free attentiveness emerges. The perception of the truth results in dropping the false. Spirituality in the act of living the reaction-free attentiveness – not knowing but direct perception. The observation is possible in a reaction-free attentiveness. Observation and understanding are in the non-dual and non-conceptual realm. The eternity, the infinity, the wholeness of life is *here* and *now*. [8]

(e) The past must be identified with memory, and future with expectations. Memory and expectation both being present facts. The ego superimposes itself on the past and the future converting it to modified [deformed] present.

(f) As long as the "me" moves, the whole paraphernalia of concepts, symbols, reactions and evaluation will travel with it; or in other words, the "me" is coupled with all thought-structures, the volitions. The "me" is the movement of time, put together by the conditioned [limited] mind. There is no growth, no freedom and no discovery. The cerebral movement, the psychological movement, the movement of the mind is absolutely irrelevant to the discovery of an alternative way of living. This paradox can only be ended by non-action of the conditioned mind, effortlessness of the mind.

(g) An alternative civilisation will have to be founded upon ending the movement of thought-time and "me" or "ego", and to live in intuitive perception. The new dynamics of human relationship is free of all violence, conflict, nervous tension and chemical pressures.

(h) To sum up: (i) we have to expose the conditioned and limited innovated-structure to direct encounter with the awareness of the scientific temper; (ii) it leads the ego and psychological imbalances to go to abeyance; (iii) it will flow to intuition which will radically transform the old innovated-structure to the objective-fact; and (iv) the objective-fact leads to commercially useful process/product, which will beat global competition. We as a nation will have to innovate to survive.

References

1. Hideki Yukawa, *Creativity and Intuition,* Tokyo: Kodansha International Ltd., 1973, pp. 123, 125, 126.
2. Erwin Schrodinger, *What is Life? & Mind and Matter,* Cambridge University Press, 1967, p. 152; and *Science and Humanism,* Cambridge University Press, 1951, pp. 4, 51.
3. Maurice H. Wilkins, *Synthesis of Science and Religion,* Mumbai: Bhaktivedanta Institute, June 1988, p. 29.
4. Voltaire, in Ilya Prigogine and Isabelle Stengers, *Order Out of Chaos,* London: Bantam Books; 1984, pp. 257.
5. Sampooran Singh, *The Holistic Education,* New Delhi: Siddharth Publications, 1987, pp. 160-161.
6. *ibid., A Scientific Guide to the Integration of Man, Society and Mankind, ibid.,* pp. 80, 84, 86.
7. Paul E. Plsek, *Creativity, Innovation, and Quality,* New Delhi: Prentice Hall of India, July 2003, Preface.
8. Vimala Thakar, *The Benediction of Being Alive,* Bookfund Vimala Thakar, The Netherlands, April 1984, Nr. 15, p. 36.

17

What is Life?

1. Introduction

Life is a mysterious wholeness; it is homogeneous, unfragmentable and dimensionless. It is immeasurable and unnamable. Life expresses itself as heterogeneous, fragmented and multidimensional. Every existence has its own characteristics. The superstructures of concepts have been created by the conditioned human mind and grafted upon reality, which results in distorted perception. What we call our living is only a chain of reactive activity. Our life is a movement of reactions, of habit patterns.

Life is self-generated, self-controlled, self-regulated and self-sustained phenomenon. Life seems to be an Intelligence, which has been unwinding, manifesting itself for millions and billions of years and yet has remained ever virgin and ever fresh.

There is One Reality, the Unmanifest-Manifested. The Active Reality and the Passive Reality are different states of the same Reality. The Active Reality is a powerful, highly directional, monochromatic [same frequency] and coherent [the wave is in one phase] beam of energy; or a state of highest frequency [higher frequency means higher energy], or the highest quantum energy potential of consciousness. The intense energy of insight expresses itself as different perceptions of the fact, the Reality.

This article explores the inner dimensions of life. It also explores the creativity which is Nature's response to our challenges, so it deals with *intuition, invention* and *discovery;* it also discusses innovation which refers to the use of Nature's response at a lower level of awareness which converts it to a commercially useful process/product.

2. Mind-Brain-Body System?

(a) The conditioned mind-brain-body system constructs the time-space matrices in quick succession. The succession leads to the concept of causation. What we call mind [with a lower case of 'm'] is a conditioned energy, moving through the body, monitored by brain, operating through the senses. It is the movement of structured framework of patterns and the dynamics of behaviour patterns, which are shared by all human beings. The time-space matrices represent the relative field of consciousness. It constructs nervous tension, chemical pressure, pain, misery, suffering and travails. All knowledge, concepts, ideals, intellectual approaches are sterile in transforming the conditioned psyche, or to bring a radical revolution in man.

(b) Erwin Schrodinger discussed two general facts; that all scientific knowledge in based on sense perception, and that non the less the scientific views of natural process formed in this way lack all sensual qualities and therefore cannot account for the latter.[1]

(c) Wilder Penfield, an eminent Canadian brain surgeon, concluded, "Our minds and our brains are not at all the same? .... it is the mind which experiences and it is the brain which records the experience"[2], and, "to suppose that consciousness or the mind has a location is a failure to understand neurophysiology".[3]

(d) All psychological imbalances – violence, conflict, fear, greed, anger, terrorism, and so on are constructs of the waking state of mind, or relative frame of reference of the mind spectrum. Violence is a construct by symbolic-dualistic frame of reference and nonviolence is a percept in the non-dual and non-conceptual frame of reference of the Mind Continuum [Mind with upper case 'M'].

(e) Albert Einstein's brain, after his death, was preserved. The eminent brain surgeons explored his brain and found that only 7-10 per cent neurons in the brain were active and the balance neurons were dormant.

(f) Candace Pert wrote, "White blood cells are bits of the brain floating around in the body".[4] Ultimately, this implies that cognition is a phenomenon that expands throughout the organism, operating through an intricate chemical network of peptides that integrates our mental, emotional and biological activities.

(g) Seeing is direct, hearing is direct, touching is direct, and thinking is indirect. If one is alert, attentive, sensitive, then when he sees a thing, the totality of the unit is perceived by the eyes; and there is direct seeing.

(h) Albert Einstein stated that we need, "objective observing and understanding" to attain the "peace and serenity". It is the non-dual and non-conceptual frame of reference of Mind Continuum that has the capacity of observing and understanding. In other words, observing and understanding requires an austerity that after perception, the subjective reactions do not come up and get mixed up with the perception. Observation is a non-reactional attentiveness. It leads to inner silence. We stand at the frontiers of the known [scientific temper] and the horizons of the unknown [non-reactional attentivity] is before us.

(i) Fred Alan Wolf wrote, "We are beginning a new age of awareness, the age of quantum consciousness, the age of conscious atom. By looking within ourselves, we may be able to solve the problems facing us on the final frontier – the frontier of the human spirit".

(j) To sum up, this is the birthright of every human being to transcend this worn out mind-brain-body system and the cerebral organ which has exhausted its faculties and capacities: memory, ideation, imagination, permutation-combination of all the faculties – it has created a parallel *computer brain* [in communion with nature] and transmitted nearly all the activities of the cerebral organ into it. This state, in common parlance, is called "Invention" or "Discovery" – it is always direct perception or direct response of nature.

3. Limitations of Objective Science

Many eminent scientists have discussed that the present objective science is leading us to self-annihilation. H.K. Dua stated, "The tragedy is that those who are clear about their vision – they are only a few in number – are finding themselves unable to take others along; and those who do not have the vision but only personal interest to guide them".[5]

Jacques Monod wrote, "We are now heading rapidly towards self-extinction, towards removing mankind from the cosmic scene .... By now the problem is not self-transcendence, but survival"[6] He added, "What we see opening before us today is an abyss of darkness"[7].

Harlow Shapley wrote, "The danger is man himself. He is his own worst enemy. He is acquiring tools and studying techniques that might solve the problem we set – the complete elimination of Homo from the planet earth.[8]

Eckhart Tolle stated, "An unprecedented shift in consciousness is indeed happening on our planet". [9]

Albert Einstein observed, "The current trends like the present bring him to unimaginable sorrow. Nothing of all that will remain but a few pitiful pages in the history books, briefly picturing to the youth of the future generations the follies of its ancestors".[10]

In 1932, after a World Disarmament Conference, in Geneva, Albert Einstein in an unusual bout of anger, exclaimed, they [politicians and statesmen] have cheated us, they have fooled us. Hundreds of millions of people in Europe and in America; billions of people have been and are being cheated, traded, and ticked out of their lives, health and well being".[11]

Wilkins wrote, "I agree with Einstein that the sort of scientific education we have now has produced a narrow-minded way of thinking among scientists, so that they give no proper attention to the moral and spiritual dimensions.[12]

The above quotes are equally valid for psychological mode of mind, or relative frame of reference of mind spectrum today. In principle, man is deceiving himself. He has become a puppet, a mechanical golem. He has last sensitivity, he has lowered his level of perception and is basically engaged in gathering information [knowledge] experience memory, ideals, motives and so on. He has started living in a synthetic world which is totally divorced from Reality. These quotes are equally valid for innovated structures.

4. A Critique on Innovation and Scientific Temper

When the non-dual and non-conceptual frame of reference of Mind Continuum decides to follow technology – which is systematic orchestration of all knowledge and experience - to lead something practical and commercially useful; the mind *slides down* [quantum fall] from scientific temper to psychological mode. It is now governed by dualism, so it often leads people to destruction. The conditioned mind deals with materialism, and it deals with thought-ego.

If a scientist with scientific temper moves through the conditioned mind, through the thought-ego, with the idea of converting "innovation" to a commercially useful process/ product. The invention deals with truth and non-reactional attentivity; and innovation deals with false and reactions only. The innovation discusses the psychological or relative frame of reference of mind spectrum; it observes the organized collective thoughts of the whole human race and I am a part of that. I see the coupled thought-ego, which destroys the innovation process.

When you conduct an enquiry, with each step there is a movement away from the false in our daily life, in our daily relationships. When I learn observation, the reaction-free attentiveness comes out to be there as a steady flame in my consciousness. This heals the innovated structure and the scientist is there to beat global competition.

5. Intelligence

Intellect, or the mind and its whole contrivance, is the creation of thought-ego structure. Intelligence is not a move of intellect, and it takes in its sweep the wholeness of life, and aware of itself as a part of the wholeness. The movement of Intelligence is awareness of the wholeness

The energy of Intelligence has nothing to do with our brain, our sensual system, our body; it has nothing to do with inheritance, conditioning, knowledge. It is an energy born of silence. Silence and emptiness are the content of that energy.

The awareness of the whole stimulates love, compassion, a tenderness of the heart and concern for all. The responses are spontaneous – there is no delay time between perception and response.

If observation, understanding and learning is continued for a long chronological time, the *"Pure Universal Consciousness"* expresses Itself at the highest quantum energy potential of Consciousness, or the Consciousness express Itself to Itself. This state, in common terminology, is called "Discovery".

The world situation is grave, we need a drastic social action of exploring alternative ways of living and new dimensions of consciousness. We have to explore the psychic avenues and new dynamics of human relationships.

6. What is Life?

(a) Man is an amphilian who lives alternately in two worlds – the inner dimension, which is eternal, infinite, immutable and unknown; and the outer dimension, which is transient, short-lived, mutable, and the known. The inner dimension is leading us to perfect freedom and peace; while the outer dimension leads mankind to self-annihilations.

(b) We find ourselves in a universe that leads life and possesses the very particular properties that make that possible.

(c) The intuition, invention and discovery belong to creativity, that deals with truth and non-reactional attentivitiy. The innovation, or the psychological mode of mind deals with false and reactions only. The innovation is a construct of the conditioned psyche.

(d) All healing is by the higher potential of consciousness. Consciousness or Intelligence is the only healer.

(e) It is a benediction of nature where thought-ego go abeyance and Intelligence gets activised. Everything is governed by immutable laws of nature.

(f) Albert Einstein wrote, "What is truly valuable in our battle of life is not the nation, I should say, but the creative and impressionable individuality, the personality – he who produces the noble and sublime while the common herd remains dull in thought and insensible in feeling".

7. Concluding Comment

The psychological mode of mind is dualistic, repetitive and mechanistic. It has no newness or no originality.

Transformation comes about as a happening in the emptiness of the mind, or in scientific temper. Transformation comes through *intuition, invention* and *discovery.*

It is a benediction that comes about when thought-ego and mind go to abeyance. Man is born to be free, to live in peace, in imperturbable peace within.

As a metaphor we say that the intellect [or reason] is just a bubble of water in the ocean, while the "inner awareness" called creativity – is just like a borderless, unfathomable, limitless ocean of life. Everything is governed by immutable laws of Nature; and everything is a necessary effect. Man's survival and existence are embedded in the creative process of Nature. Man has to jettison intellect and imbibe intuition.

India was once the "Light of Asia". She has now to fulfill her historic role of being the Light of the World. India has been called upon *here* and *now* to save Life in the world from extinction. Will she wake up from deep slumber and ***"Lead Kindly Light Amid the Encircling Gloom".*** We have the privilege and the responsibility.

References

1. M.M. Sharma, *Defence Laboratory,* Jodhpur, July 14, 2004.
2. Erwin Schrodinger, *What is Life? & Mind and Matter,* Cambridge University Press, 1967, p. 177.
3. Wilder Pemfield, in Deepak Chopra, *How to Know God,* London: Rider, 2000, pp. 215, 217; and *ibid.,* in Fred Alan Wolf, *Taking the Quantum Leap,* 1981, p. 248.
4. Candace Pert, in Fritjof Capra, *The Web of Life,* Flamingo, 1996, p. 277.
5. H.K. Dua, Editor-in-Chief, *The Tribune,* August 15, 2006.
6. George Wald [Nobel Prize in Medicine, 1967], in *The Evolution of Consciousness,* New Delhi: National Publishing House, 1983, pp. 13, 14.

7. Jacques Monod [Nobel Prize Winning Biologist], in *Chance and Necessity,* New York: Vintage Books, 1972, p. 171; in *Synthesis of Science and Religion,* Mumbai: The Bhaktivedanta Institute, June 1988, pp. 126, 127.
8. Harlow Shapley, *Of Stars and Men,* London: Elek Books, 1958, p. 144.
9. Eckhart Tolle, *The Power of Now,* Hodder 2004, Preface [xiii].
10. Albert Einstein, *Out of My Later Years,* Greenwood Press, Connectual. Ed. 1970, p. 8.
11. *ibid.*, in *Bridging Science and Spirituality,* Mount Abu, Brahma Kumaris World Spiritual University, 1988, pp. 67-68; & *The Pilgrimage to the Third Millennium,* Chandigarh: Faith Publishers, 1995, p. 16.
12. Maurice H. Wilkins [Nobel Prize in Medicine, 1962], in *Synthesis of Science and Religion, op. cit.,* p. 35.

18

The Battle of Human Psychological Imbalances

1. Introduction

Man has an evolutionary origin. He evolved as a unicellular plant at about 3.1 billion (10^9) years ago, as Homonid [man] at about 3.6 million years ago, and as Homo sapiens at about 2.5 million years ago. Fritjof Capra wrote, "the anatomical evolution of human nature was virtually completed some fifty thousand years ago. Since then the human body and brain have remained essentially the same in structure and size. On the other hand, the conditions of life have changed profoundly during this period and continue to change at a rapid pace".[1] He added, "from subatomic particles to galaxies and from bacteria to human beings – are manifestations of the universe's self-organising dynamics, which we have identified with the cosmic mind".

Reason is a precious value thrown up by evolution and the source of much human progress in culture and civilisation. Since early times, man has been observing, comparing, and seeking to classify his observations. He recalled the past memory and there is coherent superpositions of different webs on the fact [present] resulting into distorted perception and modified memory. He jettisoned the reality *and started living in the false, which is man-made world. The distorted perception leads to nervous tension, chemical pressure, violence, conflict, fear, terrorism, and so on. It is leading mankind to self-annihilation.*

The "anatomical evolution of human nature" appears to be related to psychosocial evolution of man. Many eminent scientists-philosophers-mystics have opined that "an unprecedented shift is consciousness in happening on our planet". [2]

This article discusses the battle of human psychological imbalances for the past 2.5 million years, but man has failed so far to resolve the psychological imbalances. It also explores symbiosis of creativity, science, innovation and education. It discusses the causes of limitations of science and innovation. It also presents holistic education which is bound to result in total transformation of our society into a new culture, a new civilisation.

2. What is Consciousness?

Fritjof Capra, an eminent physicist, wrote, "By Consciousness, I mean the state of mind characterized by self-awareness, where one is aware that he is aware".[3] Eugene Wigner, Nobel prize in physics, 1963, wrote, "Self-identity demands consciousness. You would not have identity without consciousness. Yes, consciousness creates identity. I know that I am I. And it is my consciousness which knows, which tells me that".[4] Wigner believes that "Our consciousness alters the world itself because it alters how we appraise the future. We experience the world the way we do because we choose to experience it that way". [5] In Everett's view, "Consciousness in the quantum connection between the numerous layers making up the quantum machine. In other words, we are quantum mechanical golems". [6]

It is the mind and the brain, which are independent; or in other words, our minds and our brains are not at all the same. It is the mind, which experiences and it is the brain which records the experience.[7] Mind is rather the all-inclusive yet dimensionless reality of which every level represents an illusory deviation. Mind is everywhere and every when.[8] The mind has no location.

3. Creativity and Innovation

Creativity is defined by Hideki Yukawa as, "the discovery of a truth in nature hitherto unknown to man .... Science is a manifestation of man's efforts to open up the unknown world that lies before him, to discover new possibilities"[9]. Creativity is direct nature's response, or direct perception. It deals with *intuition, invention* and *discovery.*

When a scientist wants to use his "invention" for technology for something commercially useful, he is caught in network of thought-ego, dualism and psychological imbalances, which destroys innovation.

The intellect and reason cannot work without measurement, without psychological time, time is the construct of thought. Ego is a concept, it has no factual reality. So thought-ego go together.

The answer to this paradox is that the scientist will have to re-explore the whole system without any psychological imbalance, or motive, or acquisition or violence, and this exploration will regenerate the innovated model. This will lead to scientific temper, which implies that the non-dual and non-conceptual frame of reference of Mind Continuum is regenerated. The "intuition and invention" gives the true objective and subjective values of life.

4. Holistic Education

The eminent philosophers and educationists – Swami Vivekananda, Sri Aurobindo, Radhakrishnan, Zakir Hussain,

Mahatma Gandhi and so on – opined that "no education worthy of the name can be conceived as something external, having no relationship with man's inner most being and becoming", and task of education is "inward transformation, or inward flowering and character building". The Divine Mother quoted Sri Aurobindo and wrote, "Man is only a transitional being living in a mental consciousness, but with the possibility of acquiring a new consciousness, the truth consciousness, and capable of living a life perfectly harmonious, good and beautiful, happy and fully conscious ...".[10] Man can achieve the supramental ideal through holistic education. The foundation of education is not knowledge [outer] but intuition [inner].

Knowledge: The movement of knowing is a cerebral activity. Knowledge constructs dualism [subject-object bifurcation]. It is a construct of the human mind, it is a man-made social cosmos. The activity of knowing is not an activity of understanding. The true aim of education is to transcend knowing [knowledge] and imbibe perception; or in other words, to make a quantum jump from dualistic frame to non-dual and non-conceptual frame of reference of Mind Continuum. This suggests that the present educational structure in India leads us to self-annihilation. It is only holistic education that helps man to grow from knowledge to understanding, from confrontation to cooperation, from violence to nonviolence, from rationality to intuition and intellect to Intelligence.

The Wholeness of Life: Maurice H. Wilkins wrote, "It is also a question of how science is developed through educational institutions and through institutionalized science. I agree with Einstein that the sort of scientific education we have now has produced a narrow-minded way of thinking amongst, scientists, so that they give no proper attention to the moral and psychological dimensions .... This is where trying to make some kind of synthesis is very important. Science as it is today doesn't tell us very much about the real nature of human beings.[11]

Roger Sperry spoke of the importance of a neglected area of science, namely the study of the human subjective experience.[12] John Eccles emphasizes the need to deal scientifically with the neglected area of inner conscious awareness.[13]

The holistic education is a new experiment at a collective level which is bound to unify, to coalesce to a new movement, to a new understanding, that is bound to result in total transformation of our society into a new culture, a new civilisation. The rising of a new culture is basically a higher perception at a higher quantum energy potential of consciousness, or a new dimension of consciousness. It is a quantum jump in consciousness – from knowledge to understanding, experience to perception, reason to intuition, and intellect to Intelligence. The problem facing mankind is to become aware with a new dimension of consciousness, which will generate a new dimension of relationship.[14]

The Action Plan is as under:

(a) To educate the educator is to have him understand himself. This is one of the most difficult undertakings, because most of us are already crystallized within a system of thought or a pattern of action. The teachers training programme is *how to think* and not *what to think.*

(b) We have to introduce psychodynamics of mind-brain-body system in science faculty and transpersonal psychology in art faculty as a part of educational curriculum at University/College/School level.

(c) The principle governing restructuring of education is fundamentally based to assist nature for an evolution of life. The evolution of mankind can only happen when there is understanding and spontaneity; dynamic interplay of intuition and action; and awareness of the transcendent dimension of life.

(d) There will be need to create some infrastructure for conducting courses. There is need to write books for guidance of teachers and students, and for different levels.

(e) The international terrorism cannot be ended through intellectual approaches, but by mutation of the human psyche through holistic education. Man has to realize *here* and *now* that the subjective dimension and the objective dimension are a single dynamic flow, a unitary movement. This implies that man has to understand both the subjective and the objective dimension of life or the wholeness of life, and live in the light of that feelingful understanding.

(f) We intuit that the holistic education [objective observing and understanding] is bound to enhance [about three times] the R & D potential of innovation. This will also guarantee the human survival and excellence.

5. Our Present Pursuits are Leading us to Self-Annihilation

Many Nobel laureates, especially in physics, are quoted as under:

(a) Harlow Shapley, an eminent scientist, wrote, "The danger is man himself. He is his own worst enemy. He is acquiring tools and studying techniques that might solve the problem we set – the complete elimination of Homo from the planet earth".[15]

(b) George Wald wrote, "By now the problem is not transcendence but survival. If we are to survive there must be fundamental changes in direction, not only organizational, but in our ways of thinking. We have to think different thoughts and want different things.[16]

(c) Huston Smith quoted, "If anything characterized modernity, it is loss of faith in transcendence, in a reality that encompasses but surpasses our quotidian affairs".[17] He added,

"Science in the process of showering us with material benefits has all erased transcendence from our reality map".

(d) Jacques Monod, a Nobel prize winning biologist, wrote, "The nineteenth-century scientism saw leading infallibly upwards to an empyrean noon hour for mankind, whereas what we see opening before us today is an abyss of darkness".[18]

(e) Hideki Yukawa wrote, "The development of science, far from encouraging the formation and development of an integrated view of man in his many different aspect, has tended rather to split and destroy the view.... The danger that all these trends will combine to destroy the unity of man's various aspects and lead eventually to the loss even of his humanity cannot be denied. It is equally undeniable, I believe, that such a split would deprive man of his happiness".[19]

(f) P. Krishna, an eminent physicist, in a lecture in USA said, "Such a mutation is necessary if a new culture is to come about. But if we go on in this way, I am afraid we are going towards self-annihilation and not much time is left now".[20]

(g) In 1932, after a World Disarmament Conference, in Geneva, Albert Einstein in an unusual bout of anger, exclaimed, they [politicians and statesmen] have cheated us, they have fooled us. Hundreds of millions of people in Europe and in America; billions of people have been and are being cheated, traded, and ticked out of their lives, health and well being".[21]

(h) Albert Einstein, out of his later years, wrote, "Current trends like the present bring him unimaginable sorrow. Nothing of all will remain but a few pitiful pages in the history books, briefly picturing to the youth of futhre generations of the follies of the ancestor".[22]

(i) Karl Sagan, an eminent astrophysicist, said, "What a waste it would be after four billion tortuous years of evolution if dominant organism contrived its own self-destruction. We are the first species to have devised the means"[23]

(j) Eckhart Tolle wrote, that there, "the urgent need for a radical change in human consciousness if humanity is not to destroy both itself and the planet .... You are here to enable the divine purpose of the universe to unfold. This is how important you are".

6. Resolution of Human Paradox

(a) George Wald, Nobel Prize in Medicine, 1967, wrote, "All the mystics through the ages have agreed that to achieve wisdom one must quiet the brain, stop using it as a tool, put aside knowledge and learning, try indeed to free one's consciousness, empty one's mind, and in quiet see what enters". What one hopes is direct perception of Nature.

(b) Hideki Yukawa, a Japanese Nobel laureate in Physics, was, "A great physicist and a great human being is deeply troubled by the predicament of mankind endowed with too much knowledge and power ....".[24] He wrote, "To save ourselves, by such efforts, from the danger of a loss of, or a split in, our humanity in the world of the future – this, I believe, is the course that man must take from now on. This, it is my personal conviction, is true wisdom for modern man".

(c) Kishore Gandhi wrote, "There are many thinkers, such as Teilhard de Chardin, Tagore, Lecomte du Nouy and Julian Huxley, who are speaking their insights. Both Teilhard de Chardin and Sri Aurobindo foresee the coming of a new race and consciousness, the Noosphere and the Supermind. Julian Huxley thinks that the great opportunity today will be found in applying scientific methods to the exploration of man's inner life".[25] We intuit that an unprecedented transformation [mutation] is happening on our planet, which is a gigantic process and enacted by Nature.

7. Intelligence

Movement of knowing is a cerebral activity. Knowledge constructs dualism – subject and object relationship. Knowledge is a construct of the human mind – man-made cosmos. The activity of knowing is not an activity of understanding. When the observer goes to abeyance, understanding is born. Understanding happens when there is non-dual and non-conceptual frame of reference of Mind Continuum. Intelligence leads to self-awareness, to pure perception, to understand and action. Intelligence is direct response of nature; it is unfettered conditioned energy. Intelligence is the highest quantum energy potential of Universal Consciousness and is independent. The human being seems to be born with the faculty of self-awareness and understanding. That understanding which does not know has a perceptive sensitivity.

Silence or stillness or solitude is a part of life. The non-dual and non-conceptual frame of reference of Mind Continuum is synergistic to the dualistic waking state of consciousness, so the mind makes a quantum jump from psychological state of mind to the state of awareness. The state of awareness expresses itself as intuition [insight], so it is governed by thought-free consciousness. Meditation is a state of consciousness which has perceptive intelligence. The content is not thought, information [knowledge], experience, memory. *The content is the emptiness of silence which is full of Intelligence.*

8. Psychological Imbalances [Violence]

(a) Ending of Violence : The mind started splitting into the subject [inner world] and the object [outer world] at about 2.5 million years ago. This led to coherent superpositions of different webs; and both got distorted. The whole system was open to the environment and closed to the Absolutes, the Nature. This gave birth to dualism, a conditioning of the psyche, which is leading mankind to self-annihilation.

The violence is a construct by symbolic-dualistic frame of reference of the mind spectrum, whereas non-violence is a perception by non-dual and non-conceptual frame of reference of Mind Continuum. The non-violence is synergistic or at higher quantum energy potential of Consciousness.

Violence is an indication of psychological immaturity. When the non-symbolic and non-dualistic frame of reference sees the cause of psychological immaturity, all violence goes to abeyance and non-violence self-expresses itself.

Succeeding generations have found themselves confronted by the inner conditioning of mind – say, Plato more than 2,000 years ago; Herbert Spencer at 1961; Swami Vivekananda about a century ago; C. Rajagopalachari in 1922; Mahatma M.K. Gandhi, Lord Wavell; R.D. Laing; Roger Lanphear; V.R. Krishna Iyer; and Albert Einstein.[26]

This equally is applicable to other psychological imbalances[27] – such as terrorism, fear, acquisition. It is hierarchical level of Consciousness that heals.

9. Concluding Comment

Modern science has brought man on the control panel of mutation or transformation of human psyche. Scientists have discovered Life Field or Universal Consciousness, which is the substratum of all existences, and which is at a very high frequency (higher frequency means higher energy) throughout the cosmos. We are on the threshold of extending this energy into human dimension.

It is the integral dimension of Consciousness that can resolve all challenges facing mankind. It is to heal the diseased mind-brain-body system. It is an open call to express higher hierarchical level of expression of consciousness eloquently through education for a global society and our daily living. This is the only guarantee to human survival and excellence.

If we enquire into the innovated structure, it leads to non-action of the psyche [past], which is cessation of the mental activity. It is observing things as they are, understanding them as they are, It blossoms to the silence, the inner relaxation. It guides to the movements of the wholeness, so it regenerates [heals] the innovated model. Holistic education always bestows higher awareness.

The challenge is addressed not only to the nature of objective science, innovation and education, but to the nature and quality of consciousness [mind] and life. We have to understand the limitations of the intellect and reason; to be open to observation and understanding; and to imbibe freedom and order. We require open-mindedness, non-authoritarian approach to life. This awakens higher awareness, which leads to Intelligence, which has the capacity to resolve all violence, conflict, terrorism and psychological imbalances. Intelligence is also called Truth-Consciousness, and is capable of guiding life perfectly harmonious, good and beautiful, happy and fully conscious, and aware of the wholeness of life.

Reference

1. Fritjof Capra, *The Turning Point,* London: Wildwood House, 1982, pp. 325, 324.
2. Eckhart Tolle, *The Powers of Now,* Hodder, 2004, Preface [xiii, xv, vii].
3. Fritjof Capra, in *Synthesis of Science and Religion,* Mumbai: The Bhaktivedanta Institute, June 1988, p. 275.
4. Eugene Wiger, in *ibid.,* p. 262
5. *Ibid.,* in Fred Alan Wolf, *Taking the Quantum Leap,* San Francisco: Harper and Row, 1981, *ibid.,* p. 211.
6. Everett, in Fred Alan Wolf, *ibid.,* p. 211.
7. Wilder Penfield, in Deepak Chopra, *How to Know God,* London, Rider, 2000, pp. 215, 217.
8. Ken Wilber, *The Spectrum of Consciousness,* USA; The Theosophical Publishing House, 1982, (Quest Book), pp. 298, 299.
9. Hideki Yukawa, *Creativity and Intuition,* Tokyo: Kodansha International Ltd., 1975, pp. 113, 205.

10. The Mother, *An Introduction to True Spirituality, Words of Shri Aurobindo and Mother,* All India Magazine, Shri Aurobindo Society, Pandicherry, Nov. 1997, p. 9.
11. Maurice H. Wilkins, *Synthesis of Science and Religion, ibid.,* June 1988, pp. 29-38.
12. Roger Sperry, *Synthesis of Science and Religion, op, cit.,* p. 199
13. John Eccles, *ibid.,* p. 199.
14. Sampooran Singh, Kanwaljit Kaur and Paramjit Singh, *Global Education and Human Survival,* 2002, pp. 68-69.
15. Harlow Shapley, *Of Stars and Men,* London: Elek Books, 1958, p. 144.
16. George Wald, in Kishore Gandhi, *The Evolution of Consciousness,* New Delhi: National Publishing House, 1983, p. 14.
17. Quoted by Huston Smith, *The Chronicle of Higher Education,* Jan. 9, 1978, in *Synthesis of Science and Religion, op. cit.,* pp. 121, 122.
18. Jacques Monod, *Chance and Necessity,* New York: Vintage Books, 1972, p. 171; *in Synthesis of Science and Religion, op. cit.,* p. 127.
19. Hideki Yukawa, *op. cit.,* pp. 204, 205, 206.
20. P. Krishna, *Education, Science and Spirituality,* The University of California and Santa Barbara, April 3, 1993, p. 12.
21. Albert Einstein, in *Bridging Science and Spirituality,* Mount Abu, Brahma Kumaris World Spiritual University, 1988, pp. 67-68; & *The Pilgrimage to the Third Millennium,* Chandigarh: Faith Publishers, 1995, p. 16.
22. *ibid., I Believe: Nineteen Personal Philosophies,* London: Unwin Books, 1962, p. 31.
23. Carl Sagan, in *Science and Spirituality: Three,* Chandigarh: Faith Publishers, 1988, p. 50.
24. Hideki Yukawa, *op. cit,* cover Jacket.
25. Kishore Gandhi in *The Odyssey of Science, Culture and Consciousness,* New Delhi: Abhinav Publishers, 1990, pp. 40-41.

26. Sampooran Singh, The Survival of Mankind and Quantum Jump from Violence to Peace, New Delhi: Books India International, October 1999, pp 223 – 226.
27. Sampporan Singh, Mrs. Kanwaljit Kaur, Paramjit Singh, The Masterplan Paradigm for Human Survival and Excellence, Chandigarh: Faith Publishers, February 1998, pp xxiv + 264.

Questions and Answers

Q. What is Spirituality?

Ans. There are two worlds – the microcosm [the internal], the macrocosm [the external]. Physics has introduced the sub-microcosm [the world of infinity small, sub-atomic world]. The world spirituality is a science which deals with the wholeness of life – the macrocosm, the microcosm, the sub-microcosm and the individual as its field of exploration. The science of spirituality begins with the wholeness of life and then sees the particular in the light of the wholeness. In spiritual science, the emphasis is upon the act of perception, the quality of instruments employed for perception, and the quality of consciousness behind the perception.

Both science and spirituality follow the law of non-authoritarian approach – an approach not conditioned by the authority of the past. What is necessary is the exploration on behalf of the human race.

When the Subject Cognizance is in direct communion with nature, or in direct perception, the nature of the spiritual, energy flows to the object. The subject of cognizance is in non-dual and non-conceptual realm, there is no duality. The spiritual energy is beyond the energy of thought-ego, beyond the sentiments, emotions and other psychological imbalances, beyond the emptiness or silence of the mind.

Objective science becomes fulfilling when it imbibes the subjective division. Subjective science and spirituality are complementary. Both deal with the wholeness of life. This helps to evolve a complete civilisation for which the world is ripe and waiting.

Q. Please Discuss the Role of Science in Development of Human Personality?

Ans. (a) Ilya Prigogine and Isabelle Stengers wrote, "The dialogue with nature isolated man from nature instead of being closer to it. A triumph of human reason turned into a sad truth. It seemed that science debased everything it touched".[1]

(b) Hideki Yukawa wrote, "The development of science, far from encouraging the formation and development of an integrated view of man in his many different aspects, has tended rather to split and destroy that view".[2]

(c) Erwin Schrodinger wrote, "If you have to face the body of a deceased friend whom you surely miss, is it not something to realize that this body has never been the seat of his personality but only symbolically 'for practical references'?".[3] It is evident that the human personality is not his body (or age) but the level of "inner conscious awareness".

(d) Hideki Yukawa wrote, "Man has things [memory] that are unknown even to himself". This is past memory of millions of years. So the human personality is determined by his past memory, which is his level of consciousness. But the past psychological memory in always an illusion and it does not represent his real personality.

(e) Hideki Yukawa emphasised that both the Greek and Chinese [and India], "always presupposed the presence of the power of intuition",[4] – the importance of inspiration in scientific perception. Max Born stated at the third Pugwash Conference, "Clever, rational ways of thinking are not enough. The danger of mass slaughter .... can only be overcome by moral conviction, by a determination to replace national prides and prejudices with human love".[5]

To sum up, the human personality is his level of "inner conscious awareness", or in other words, "the hierarchical level of consciousness *here* and *now*". The Consciousness is nature's response in the human mind-brain-body system and not intellectually or sensory response, but man has to learn "objective observing and understanding" at every moment. Man has to learn to transcend the intellect and reason [man-made cosmos] and imbibe, "*intuition, invention* and *discovery*".

Man urgently needs exploration of ***"Scientific Research and Temper"*** and ***"Creativity and Innovation"***. The aim of this exploration is to initiate an open dialogue between scientists-philosophers-mystics, working in the field of "Cognitive Science", or "Physics of Consciousness", as the "Truth of Nature and Man". It is the higher hierarchical level of Consciousness that can bring mutation in the human psyche and lead mankind to survival and excellence.

References

1. Ilya Prigogine and Isabelle Stengers, *ibid.,* p. 6.
2. Hideki Yukawa, *op. cit.,* p. 204.
3. Erwin Schrodinger, *What is Life? & Mind and Matter,* Cambridge University Press, 1967, p. 134.
4. Hideki Yukawa, *op. cit.,* pp. 101, 206.
5. Max Born, in Hideki Yukawa, *op. cit.,* pp. 189-190.

19

The Survival of Mankind?

1. Where is Objective Science Going?

The civilisation in which we live today is the product of the human mind, known as modern science. Science aims at nothing but making true and adequate statements about its objects. The scientist only imposes two things, namely truth and sincerity, imposes upon himself and upon other scientists. The first is *pure science,* which tries earnestly to *understand* the truth of nature through a dispassionate enquiry; and the second is *applied science* [technology] in which the truth discovered by pure science flows as something commercially useful as process/product. This in common parlance is called innovation.

Erwin Schrodinger proposed *"The Principle of Objectivation":* He wrote, "Without being aware of it and without being rigorously systematic about it, we exclude the Subject of Cognizance from the domain of nature that we endeavor to understand". He added, "Mind has erected the objective outside world of the natural philosopher out of its own stuff. Mind could not cope with this gigantic task otherwise than by the simplifying device of excluding itself – withdrawing from its conceptual creation. Hence the latter does not contain its creator.[1]

Charles Sherrington wrote, "The material world has only been constructed at the price of taking the self, that is mind, out of it, removing it; mind is not part of it; obviously, therefore, it can neither act on it, or be acted on by any of its parts".[2] C.G. Jung wrote, "The flood of external objects of cognizance has made the subject of all cognizance withdraw to the background, often to apparent non-existence".[3] B. Spinoza stated, "Neither can the body determine the mind to think, nor the mind determine the body to motion or rest or anything else."[4] This leads us to an antinomy that which everything is a construct of the mind, yet "the conscious mind itself remains a stranger within that construct". We intuit that the exclusion of the Subject of Cognizance is basically due to a sudden encounter of the "intense energy of insight" which results in enhancing the quantum energy potential of the object. Maurice H. Wilkins wrote, "The whole question of moral, spiritual, and other dimensions is normally pushed out of science. So, something in the nature of science often does lead people towards destruction". He added, "The world view engendered by science is in many ways a very undesirable one because it focuses attention on the material aspects of life and on technical solutions to human problems. I think this is very bad for human attitudes .... Science is a human activity. The question of good and bad, of human virtue, relates to every human activity. How therefore can science, being a human activity, be separated from all virtues and values?[5]

Schrodinger discussed two facts (a) that all scientific knowledge is based on sense perception, and (b) that none the less the scientific views of natural processes formed in this way lack all sensual qualities and therefore cannot account for the latter.[6] All great discoveries have arisen from the realm of the timelessness, so scientific enquiry is not a cerebral activity and gathering information, but a quest of the Timelessness, the Divinity, the Life Field. A selfless, unfaltering devotion to truth and use of knowledge for man's uplift – higher hierarchical expression of Consciousness – are the only vocation of a scientist.

Schrodinger was astonished at finding "our world picture 'colourless, cold, mute'. Colour and sound, hot and cold are our immediate sensations; small wonder that they are lacking in a world model which we have removed our own mental person .... The physical world picture lacks all the sensual qualities that go to make up the Subject of Cognizance. The model in colourless and soundless and unpalpable".[7] John Eccles, British neurologist perceives the sensory illusions with one startling but irrefutable assertion, "I want you to realize that there is no colour in the natural world and no sound – nothing of this kind, no textures, no patterns, no beauty, no scent ....".[8] In short, none of the objective facts upon which we usually have our reality is fundamentally valid. We intuit that all the objective facts are constructs of the inner conscious awareness, the subjective aspects of the mind. This gives evidence that the mind has both objective and subjective dimensions. This leads objective science to the objective and subjective dynamics. Schrodinger wrote, "Scientific attitude would have to be rebuilt, science must be made anew. Care in needed."[9] We have to restructure our science.

Roger Sperry spoke of the importance of a neglected area of science, namely the study of the human subjective experience.[10] John Eccles emphasizes the need to deal scientifically with the neglected area of inner conscious awareness.[11]

Max Planck stated, "It is a fact that there is a point, one single point in the immeasurable expanse of mind and matter, where science and therefore every causal method of research is inapplicable, not only on practical grounds, but also on logical grounds, and will always remain inapplicable. This is the point of [our] individual awareness".[12] Edwin Hubble stated, "The world of pure values, that world which science cannot enter, has no concern whatsoever with probable knowledge. There finality – eternal, ultimate truth – is earnestly sought.[13]

2. What is Consciousness?

Consciousness performs a dual role in the universe. In the world of the quantum, it is both the awareness and the creation of information and experience. It is the *being* and the *knower* of experience. With the stroke of the twentieth-century quantum eraser, the dividing line between ontology [theory of *being*] and epistemology [theory of *knowing*] is rubbed out.[14] The fusion of ontology, epistemology and psychology is beginning to emerge.

Fritjof Capra, an eminent physicist, wrote, "By Consciousness, I mean the state of mind characterized by self-awareness, where one is aware that he is aware".[15] Eugene Wigner, Nobel prize in physics, 1963, wrote, "Self-identity demands consciousness. You would not have identity without consciousness. Yes, consciousness creates identity. I know that I am I. And it is my consciousness which knows, which tells me that".[16] Wigner believes that "Our consciousness alters the world itself because it alters how we appraise the future. We experience the world the way we do because we choose to experience it that way".[17] In Everett's view, "Consciousness in the quantum connection between the numerous layers making up the quantum machine. In other words, we are quantum mechanical golems".[18]

It is the mind and the brain which are independent; or in other words, our minds and our brains are not at all the same. It is the mind which experiences and it is the brain which records the experience.[19] Mind is rather the all-inclusive yet dimensionless reality of which every level represents an illusory deviation. Mind is everywhere and everywhen.[20] The mind has no location.

3. What is Mind-Brain-Body System?

Fritjof Capra wrote, "The anatomical evolution of human nature was virtually completed some fifty thousand years ago. Since then the human body and brain have remained essentially the same in structure and size. On the other hand, the conditions of life have changed profoundly during this period and continue to change at a rapid pace".[21] He added, "from subatomic particles to galaxies and from bacteria to human beings – are manifestations of the universe's self-organising dynamics, which we have identified with the cosmic mind".

Psychological Mode of Mind: Since early times, man has been observing, comparing, and seeking to classify his observations. He recalled the past memory and there is coherent superpositions of different webs on the fact [present] resulting into distorted perception and modified memory. This may be written as:

Pure Perception	+	Recalled Memory	→	Distorted Perception	+	Modified Memory

This is commonly called the psychological mode of mind, or the relative frame of reference. It is invariably dualistic and has no newness. It is repetitive and mechanistic. He jettisoned the *reality* and started living in the false, which is man-made world. The distorted perception leads to nervous tension, chemical pressure, violence, conflict, fear, terrorism, and so on. It is leading mankind to self-annihilation.

The conditioned mind-brain-body system constructs the time-space matrices in quick succession. The succession leads to the concept of causation. What we call mind [with a lower case of 'm'] is a conditioned energy, moving through the body, monitored by brain, operating through the senses. It is the movement of structured framework of patterns and the dynamics of behaviour patterns which are shared by all human beings. The time-space matrices represent the relative field of consciousness. It constructs nervous tension, chemical pressure, pain, misery, suffering and travails. All knowledge, concepts, ideals, intellectual approaches are sterile in transforming the conditioned psyche, or to bring a radical revolution in man.

Quantum Jump to Scientific Temper: (a) We intuit that "anatomical evoluation of human nature" initiated a new evolution – called psychosocial evolution – where the primary motivation became fulfillment and evolution become conscious and goal oriented. This happens when there is "objective observing and understanding". In the words of Albert Einstein, The dualistic frame of reference makes a quantum jump to non-dual and non-conceptual frame; this is called scientific temper.

(b) Wilder Penfield, an eminent Canadian brain surgeon, concluded, "Our minds and our brains are not at all the same? .... it is the mind which experiences and it is the brain which records the experience"[22] and, "to suppose that consciousness or the mind has a location is a failure to understand neurophysiology".

(c) Albert Einstein's brain, after his death, was preserved. The eminent brain surgeons explored his brain and found that only 7-10 per cent neurons in the brain were active and the balance neurons were dormant.

(d) Candace Pert wrote, "White blood cells are bits of the brain floating around in the body".[23] Ultimately, this implies that cognition is a phenomenon that expands throughout the organism, operating through an intricate chemical network of peptides that integrates our mental, emotional and biological activities.

(e) Albert Einstein stated that we need, "objective observing and understanding" to attain the "peace and serenity".[24] It is the non-dual and non-conceptual frame of reference of Mind Continuum that has the capacity of observing and understanding. In other words, observing and understanding requires an austerity that after perception, the subjective reactions do not come up and get mixed up with the perception. Observation is a non-reactional attentiveness. It leads to inner silence. We stand at the frontiers of the known [scientific temper] and the horizons of the unknown [non-reactional attentivity] is before us.

(f) To sum up, this is the birthright of every human being to transcend this worn out mind-brain-body system and the cerebral organ which has exhausted its faculties and capacities: memory, ideation, imagination, permutation-combination of all the faculties – it has created a parallel *computer brain* [in communion with nature] and transmitted nearly all the activities of the cerebral organ into it. This state, in common parlance, is called "Invention" or "Discovery" – it is always direct perception or direct response of nature.

(g) As a metaphor we say that the intellect [or reason] is just a bubble of water in the ocean, while the "inner awareness" called creativity – is just like a borderless, unfathomable, limitless ocean of life. Everything is governed by immutable laws of Nature; and everything is a necessary effect. Man's survival and existence are embedded in the creative process of Nature. Man has to jettison intellect and imbibe intuition.

4. Objective Observing and Understanding

Observing: Seeing is direct, hearing is direct, touching is direct, and thinking is indirect. If one is alert, attentive, sensitive, then when he sees a thing, the totality of the unit is perceived by the eyes; and there is direct seeing.

Observing implies direct seeing. Observe yourself without any identification, without any comparison, without any condemnation, just observe without adding any recalled memory. The important thing, therefore, is to be aware without accumulating the experience which awareness brings.

Understanding: It comes only when the mind is very quiet. Stillness of mind comes only when there is no process of isolation through accumulation but complete understanding of the whole process of relationship. In stillness there is no projection of thought. Seeing the false as the false and the true as the true is transformation, because when you see something very clearly as the truth, the truth liberates. What it demands is a awareness to see life as a whole.

A dualistic frame of reference of mind spectrum can neither observe nor understand. We need a non-dual and non-conceptual frame of reference of Mind Continuum that can observe and understand. It is the lack of understanding of relationship that causes the conflict.

Relationship: It means communion without fear, freedom to understand each other, to communicate relationship is really a process of self-revelation, which is a process of self-knowledge.

Holistic Education: It is of primary significance to the transformation or mutation of the whole mind-brain-body system. It deals with the wholeness of life. It is regeneration of non-dual and non-conceptual frame of reference of Mind Continuum. Education is the foundation for the scientific research and temper. We have to restructure our educational pattern.

5. The Survival of Mankind

(a) The main characteristics of objective science are : (a) Science deals with psychological fields, and as such the whole question of moral, spiritual and other dimensions are normally pushed out; (b) "the principle of objectivation'explains that the subject of cognizance goes to abeyance and one is aware of the objective field, from fact to fact; (c) science cannot be separated from values; (d) the logical, intellectual is very important for objective science; (e) it offers answers to intellectual and technical problems; (f) it constructs all types of WMD [weapons of mass destructions], (g) it leads to violence; and (h) it guides man to self-annihilation or self-destruction.

(b) Many eminent scientists talked of human survival. George Wald, Nobel prize in medicine, 1967, wrote, "By now the problem is not transcendence but survival".[25] Jacques Monod, Nobel prize winning biologist, wrote in 1972, "We are opening before us today in an abyss of darkness."[26] Max Born, Nobel laureate in Physics, at the third Pugwash Conference wrote, "Clever, rational ways of thinking are not enough. The danger of mass slaughter ... can only be overcome by moral convictions by a determination to replace national prides and prejudices with human love."[27] Hideki Yukawa wrote, "The development of science, far from encouraging the formation of an integrated view of man in his many different aspects, has tended rather to split and destroy that view."[28] Ilya Prigogine wrote, "The dialogue with nature isolated man from nature instead of bringing him closer to it. A triumph of human reason turned into a sad truth. It seemed that science debased everything it touched.

(c) Science is the search for truth. A selfless unfaltering devotion to truth and the use of knowledge for man's upliftment are the ideals of a scientist. Truth is what all earnest scientists seek. The "principle of objectivation" proposed by Erwin Schrodinger and supported by Charles Sherrington, B. Spinoza, and C.G. Jung – that "the material world has only been constructed at the price of taking the self that is mind out of it, removing it" – rejects to deal with the subject of cognizance. Obviously objective science cannot deal with the "inner world".

(d) Roger Sperry, Nobel prize in physiology and medicine, 1981, spoke of a neglected area of science, namely the study of human subjective experience."[29] John Eccles, Nobel laureate, also talked of the "inner conscious awareness". We have to consider "the primacy of inner conscious awareness as a casual reality."[30]

To sum up, the survival of mankind is embedded in the subjective (inner) dimension of life. We have to restructure science to jettison reason and intellect, and imbibe *intuition, invention and discovery*. The human survival is embedded in the subjective dimension of man, which is called the Intelligence, the Wisdom, the Superluminal inner dimension of man. Roger Sperry made a prophetic declaration in words, "Instead of renouncing or ıgnoring consciousness, the new interpretation gives full recognition to the primacy of inner conscious awareness as a casual reality."

6. Concluding Comment

We intuit that the exclusion of the Subject of Cognizance is basically due to sudden encounter of the intense energy of insight which results in Subject of Cognizance going to abeyance and this enhances the quantum energy potential of the object. This results in pushing out the whole question of moral, spiritual and other dimensions of science.

Maurice H. Wilkins, Nobel Prize in Medicine, 1962, wrote that, "Science is a human activity. The question of good and bad, of human virtue [ethical/spiritual] relates to every human activity. How therefore can science, being a human activity, be separate from all virtues and values". Erwin Schrodinger wrote, "Scientific attitude would have to be rebuilt, science must be made anew, care is needed".

Colour and sound, hot and cold are lacking in our world model. All objective facts are constructs of the inner awareness. This gives irrefutable evidence that science must take into consideration the subjective and the objective dimensions.

The consummation of all activities of human race is to recognize and work towards human psychological growth, human fulfillment and excellence, justice, peace, freedom and bliss. The healing will be possible only when there is the transformation or mutation of the individual psyche. Man has to grow from intellect to Intelligence, from knowledge to understanding, from "successive consciousness" to "simultaneous consciousness". The psychic growth or mutation of the human race is a must before the new freedom, peace and world order emerges.

Man's survival is embedded in the creative process of Nature. We have to jettison intellect and reason which leads to self-annihilation. We have to imbibe intuition which is response of nature.

India was once the "Light of Asia". She has now to fulfil her historic role of being the Light of the World. India has been called upon *here* and *now* to save Life in the world from annihilation. Will she wake up from deep slumber and ***"Lead Kindly Light Amid the Encircling Gloom".*** We have the privilege and the responsibility.

References

1. Erwin Schrodinger, *What is Life? & Mind and Matter,* Cambridge University Press, 1967, pp. 126, 131.
2. Charles Sherrington, *Man on His Nature,* in Erwin Schrodinger, *ibid.,* p. 128.
3. C. G. Jung, in Erwin Schrodinger, *op. cit.,* p. 129.
4. B. Spinoza, *Ethics, Part-III, Prop. 2;* translated by Ewin Schrodinger, *op. cit.,* p. 131.
5. Maurice H. Wilkins, *Synthesis of Science and Religion,* Mumbai: Bhaktivedanta Institute, June 1988, pp. 29-38.
6. Erwin Schrodinger, *op. cit.,* pp. 128, 148.
7. Erwin Schrodinger, *op. cit.,* pp. 177.
8. John Eccles, in Deepak Chopra, *Ageless Body, Timeless Mind,* New York: Three Rivers Press, 1993, p. 12.
9. Erwin Schrodinger, *op. cit.,* p. 132.
10. Roger Sperry, *Synthesis of Science and Religion, op. cit.,* p. 199.
11. John Eccles, *ibid.,* p. 199.
12. Max Planck, *Synthesis of Science and Religion, op. cit.,* June 1988, p. 183.

13. Edwin Hubble, *The Nature of Science and Other Lectures, ibid,* p. 119.
14. Fred Alan Wolf, *Taking the Quantum Leap,* San Francisco, Harper and Row, 1981, p. 229.
15. Fritjof Capra, in *Synthesis of Science and Religion,* Mumbai: The Bhaktivedanta Institute, June 1988, p. 275.
16. Eugene Wiger, in *ibid.,* p. 262.
17. *Ibid.,* in Fred Alan Wolf, in *ibid.,* p. 211.
18. Everett, in Fred Alan Wolf, *ibid.,* p. 211.
19. Wilder Penfield, in Deepak Chopra, *How to Know God,* London, Rider, 2000, pp. 215, 217.
20. Ken Wilber, *The Spectrum of Consciousness,* USA; The Theosophical Publishing House, 1982, (Quest Book), pp. 298, 299.
21. Fritjof Capra, *The Turning Point,* London: Wildwood House, 1982, pp. 325, 324.
22. Wilder Penfield, in Deepak Chopra, *How to Know God,* London: Rider, 2000, pp. 215, 217; and *ibid.,* in Fred Alan Wolf, *Taking the Quantum Leap,* 1981, p. 248.
23. Candace Pert, in Fritjof Capra, *The Web of Life,* Flamingo, 1996, p. 277.
24. Albert Einstein: 14 March 1879 – 18 April 1955, Kenneth Brecher, *Nature*, Vol. 278, 15 March, 1979, pp. 215-218.
25. George Wald, in Kishore Gandhi, *The Evolution of Consciousness*; New Delhi : National Publication House, 1993, p.14.
26. Jacques Monod, in *Synthesis of Science and Religion*, Mumbai: Bhaktivedanta Institute, June 1988, p. 127.
27. Max Born, in Heidki Yukawa, *Creativity and Intuition*, Tokyo: Kodansha International Ltd., 1973, pp 189 - 190, 204.
28. Ilya Prigogine and Isabelle Stengers, in *Order Out of Chaos*, London: Bantam Books, 1984, p.6.
29. Roger Sperry, in *Synthesis of Science and Religion, op. cit.*, p.199.
30. J.C. Eccles, *ibid.*, pp. 199-200.

20

The Benediction of Nature: The Energy of Intelligence

1. Introduction

Life is a multi-dimensional phenomenon and human beings who are expressions of that phenomenon are themselves multi-dimensional creatures. Life is self-generated, self-sustained, self-controlled and self-regulated phenomenon, life seems to be an intelligence, energy of intelligence, which has been unwinding, uncovering, manifesting itself for billions of years and yet has remained ever virgin and ever fresh. Life is Infinite and Eternal, so life is an expression of Eternal Intelligence.

Life is full of love and compassion, nonviolence and non-conflict, harmony and peace. The creativity *[intuition, invention and discovery]* are natures direct perceptions, and this leads man out of prison of his own conditioned mind. This is a superhuman task as one out of ten-million people have achieved this state of the non-dual and non-conceptual frame of reference of the Mind Continuum. This is called the absolute rest frame of reference of Mind Continuum. The rest remain in dualistic and relative frame of reference of mind spectrum, which is often called man-made cosmos.

The Mind Continuum energies the Cosmic Intelligence, which takes over the charge of one's being and ends all violence and conflict, and bestows love and compassion, freedom and peace.

Human Challenges: This article discusses the existences in the man-made cosmos; the 'I' or 'me' or 'ego', the cause of human degeneration; science for the development of human personality; meditation and spirituality. The mind identified state is severely dysfunctional. It is a form of insanity. It is disharmonious and disordered. It lives in the past memory. Psychologically we go through agonies, tortures, uncertainties, despair, a great sense of sorrow.

2. Recent Natural Disasters

On March 1, 2007, a deadly twister, a tornado struck Alabama, US *[The Hindu, March 3, 2007]*. An earthquake on Indonesia's Sumatra, Island struck on March 6, 2007 killing 82 persons, destroying buildings *[The Tribune, March 7, 2007]*. Such disasters take a heavy toll of human life and property year after year.

3. Existence in the Man-Made Cosmos

From time immemorial, man had been living in the psychological mode of mind or relative frame of reference of mind spectrum. The psychodynamics of the mind-brain-body system are : (a) in Spencer Brown words, "in order to see, the mind cuts itself into at least one state which sees and at least one other state which is seen",[1] this is called subject-object dualism; (b) the seer [subject] gathers information [knowledge] and experience and records it in the brain as memory; (c) there is coherent superposition of the subject on the object resulting into distortions of both the subject and the object; and (d) we are aware of the distorted perception. It implies that man has lost the faculty of direct seeing, of direct hearing, and so on. Sullivan put it, "We cannot observe the course of nature without disturbing it",[2] or Andrade, "Observation means interference with what we are observing .... observation disturbs reality".[3] Man is aware of the distorted perception and distorted memory is recorded in the brain.

Albert Einstein stated, "I do not believe that we have any freedom at all in philosophical sense, for we act not only under external compulsion but also by inner necessity".[4] Schopenhauer's saying, "A man can surely do what he wants to do, but he cannot determine what he wills".[5] The loss of freedom in the philosophical field has made man a repetitive and mechanical golem.

4.The "I", or "me", or "Ego"

The ego is an identification with the name, it goes deeper and becomes stronger with every response. The demands, gratification, identification preserve the ego. The 'I' feels alive only in the movement – movement of acquiring or movement of reacting.

The ego does not have any objective reality. It has no factual reality at all. It has a conceptual reality. It is invariably coupled with thought-structure. When thought moves, ego emerges; so thought is a structural movement. The functions of thought are comparison, competition, acquisition, and so on.

When a thought travels in the body, it consumes nervous energy. When an emotion travels in the body it affects the chemical system. When thought goes to abeyance [based on objective observing and understanding], the education in silence starts – silence [solitude, stillness] implies remaining inwardly empty.

When there is silence, there is physical relaxation and at the same time there is flame of attentivity or sensitivity. It is non-subjective and non-objective attention. Discontinuity of the movement of information [knowledge], experience and inheritance is silence. The energy of Intelligence gets activised. All violence and conflict, psychological imbalances go to abeyance.

5. The Course of Degeneration of India

J. Krishnamurti had a meeting with Jawaharlal Nehru and Acharya Kripalani in 1931. Krishnamurti said, "If they ignored inner regeneration in the struggle for Independence, India would lose its way".[6] About 60 years have passed but India has not started "inner regeneration".

The beginning was order at some stage, man made an error of collecting experience and information [knowledge], which he stored in the brain. He then made coherent superpositions of different webs on the original order, which resulted into distortions. Man made chaos, which means disorder. We have lost that sense of blessed order, so darkness of chaos has been created by man, by man-made chaos.[7]

Krishnamurti, in 1979, stated, "India has lived all these centuries on ideas. She has to move from a life based an ideas, which are non-facts, to seeing of fact only".[8] A living in the non-fact implies to live in the psychological mode of mind, which is always relative and dualistic. It is at the lowest quantum energy potential of consciousness, which is repetitive and mechanistic.

He added that the "true religious spirit that probed, questioned and negated" lost the quantum energy potential of consciousness, or in other words, it has been wiped and destroyed in India. He stated that "a presence [Unmanifest] waited in the land. The benediction of the timeless is waiting, the ground is ready".[9] It was a prophecy that a sacred energy was alive again in the Indian soil.

Krishnamurti said, "I have negated everything that man has done. I see that all that does not lead anywhere.[10] It means that negation does not lead the mind to silence. Seeing is direct, hearing is direct, touching is direct. Direct perception is true, the present, the fact, the non-duality, the what is? We have to deny knowledge, deny everything.

The Western world, the Christian world, had a religious core that rested on faith. In India at the heart of religion was the denial of everything but 'that' [Unmanifest]. In the Christian world doubt had never been part of religious enquiry, that rested on faith.[11] In India doubt as a part of religious enquiry, has always existed. Now in that religious core, the seed is disappearing. In ancient times religion was not based on caste or ritual.

In 1981, Krishnamurti said, "Learn to Die to Yourself Completely" implies that all man-made world goes to abeyance, as this invariably gives a distorted perception and we need direct perception, which is in communion with nature. The enquiry within is infinite. You must be alone, stripped, then you can take a journey into the unknown.[12]

The art of "objective observing and understanding" awakens the non-dual and non-conceptual frame of reference of Mind Continuum, and it is called the awakened insight which probes. We have to learn observing, listening and questioning. The total perception deals with the wholeness of mankind. Observation implies to see the conditioned psyche and its functioning, listen to what is within and without, and out of this arises sensitivity, and in sensitivity there arises insight,[13] which wipes away all psychological imbalances. The individual is one who steps out of the stream. The answer to this paradox is to observe the thinker and the thought as one. This gives insight which is instantaneous. Perception of the totality is an instant perception. This bestows a perception of love and compassion. This is the essence of life.

We have to renounce thinking and verbalization, and not action. Through action you come in contact with reality. Seeing is action, moving is action. The adition leads you to sensitivity, which flows to awakened insight.

It is the mind that distorts the reality and consciousness reveals it. ***Knowledge belongs to the mind, not to consciousness.*** The mind cannot live in the *now* the *here,* became now is not a part of time. Mind cannot live in the eternal *now.* Consciousness is the highest quantum energy potential of the sacred energy. It is the consciousness that heals, that brings transformation or mutation.

Albert Einstein stated that we urgently need "Objective observation and understanding",[14] and this flows from non-dual and non-conceptual frame of reference of Mind Continuum. This energies the Cosmic Intelligence, which takes over the charge of our being. In the realm of Intelligence, all disasters, catastrophes go to abeyance and man revels in love, freedom and peace.

To sum up, man has been trying for uncountable aeons [at least 30,000 years ago] to resolve the psychological imbalances, but has utterly failed. The basic cause is that he never tried to educate his mind-brain-body system, and understand his psychological imbalances. Albert Einstein, in prophetic words, said, "Objective observing and understanding" leads to "peace and serenity".[14] Mankind has to act *here* and *now* on these words.

6. Role of Science in the Development of Human Personality

We quoted Galileo Galile[15], Voltaire[16] and Herman Wey[17] [in chapter 3] To sum up, ***"inner awareness" is essential for understanding.*** We shall always be ignorant of the relationship between the world of science and the mind which knows, perceives, and creates this science".[18]

Ilya Prigogine and Isabelle Stengers wrote, "The dialogue with nature isolated man from nature instead of being closer to it. A triumph of human reason turned into a sad truth. It seemed that science debased everything it touched".[19]

Hideki Yukawa wrote, "The development of science, far from encouraging the formation and development of an integrated view of man in his many different aspects, has tended rather to split and destroy that view".[20]

Erwin Schrodinger wrote, "If you have to face the body of a deceased friend whom you surely miss, is it not something to realize that this body has never been the seat of his personality but only symbolically 'for practical references'?".[21] It is evident that the human personality is not his body (or age) but the level of "inner conscious awareness".

Hideki Yukawa wrote, "man has things [memory in him] that are unknown even to himself".[22] This is past memory of millions of years. So the human personality is determined by his past memory, which is his level of consciousness. But the past psychological memory in always an illusion and it does not represent his real personality.

Hideki Yukawa emphasised that both the Greek and Chinese [and India], "always presupposed the presence of the power of intuition",[22] – the importance of inspiration in scientific perception. Max Born stated at the third Pugwash Conference, "Clever, rational ways of thinking are not enough. The danger of mass slaughter .... can only be overcome by moral conviction, by a determination to replace national prides and prejudices with human love".[23]

Krishnamurti explored the mind-brain-body system in depth in dialogues with David Bohm and stated:

I have a million years of experience, and it has given me a certain capacity. And I realize at the end of it all there is no relationship between me and the truth .... Your brain, and yours mind, your knowledge are valueless? I discover the truth, that all churches, prayers, books have absolutely no meaning .... There is only one thing, and that is to discover that all that I have done is useless – ashes.[24]

An answer to this paradox was provided in "Preface" to *"Within the Mind"* in words, "The central core is that man has to be a light unto himself. Truth is a process of self-discovery. No one can give the truth to another .... It is through a self-enquiry".[25]

To sum up, the human personality is his level of "inner conscious awareness", or in other words, "the hierarchical level of consciousness *here* and *now*". The Consciousness is nature's response in the human mind-brain-body system and not intellectually or sensory response, but man has to learn "objective observing and understanding" at every moment. Man has to learn to transcend the intellect and reason [man-made cosmos] and imbibe, *"intuition, invention* and *discovery"*.

Man urgently needs exploration of ***"Scientific Research and Temper"***[26] ***"Creativity and Innovation"***[27]; and ***"A Vision on Tsunamis, Earthquake, Snow and Avalanches and Global Warming"***.[28] The aim of this exploration is to initiate an open dialogue between scientists-philosophers-mystics, working in the field of "Cognitive Science", or "Physics of Consciousness", as the "Truth of Nature and Man". It is the higher hierarchical level of Consciousness that can bring mutation in the human psyche and lead mankind to survival and excellence.

7. What is Meditation?

Meditation is a way of living, an alternative way of living. The conditioned mind divides itself into the subject and the object; it is division into the "I" and the "me"; this is called dualism. Meditation is a state of consciousness in which the dualism has ended. There is only a holistic awareness. The awareness of the whole permeates into the being, percolates to the sense organs. The senses behave as the non-dual consciousness.

Meditation is a dimensional transformation from dualistic frame to non-dual frame, from the dimension of life to the dimension of the "It", the Life. It is an awareness without division, without fragmentation. It is a revolution in the content of consciousness. It is a revolution in the texture of our perception and our response.

Meditation is a quantum jump from psychological frame of reference to "Scientific Research and Temper". It is mutation in the whole being – in the behaviour, the speech, the consciousness.

When a thought travels in the body, it consumes nervous energy. When an emotion travels in the body, it affects the chemical system, it affects the health. There is a racial consciousness, which gives racial momentum of knowledge and experience. I gratified their needs and I acted upon them, which is the "me" so there is nothing like an individual ego, the "me". The momentum of thought structure is independent of your volition. There is nothing like an individual mind and there is nothing like an individual "I-consciousness".

To wash the toxins of exhaustion, they invented the technique of sitting quietly, educating the physical body to be still completely. Relaxed but alert. They educated the mind to observe without identification, without motivation, without acquisition. The perception is there, the observation is there, but there is no reaction. It is a state of no-mind. Education is a religious pursuit – there was abstinence from verbal speech, abstinence from seeing, abstinence from learning. There is no response, just the watching, that watching becomes a flame of attention. Non-subjective, non-objective attention is there. Silence is the non-action of the ego. This energies the healing energies, the energy of Intelligence, which resolves all horrors of man-made cosmos.

8. What is Spirituality?

There are two worlds – the microcosm [the internal], the macrocosm [the external]. Physics has introduced the sub-microcosm [the world of infinity small, sub-atomic world]. The word spirituality is a science, which deals with the wholeness of life – the macrocosm, the microcosm, the sub-microcosm and the individual as its field of exploration. The science of spirituality begins with the wholeness of life and then sees the particular in the light of the wholeness. In spiritual science, the emphasis is upon the act of perception, the quality of instruments employed for perception, and the quality of consciousness behind the perception.

Both science and spirituality follow the law of non-authoritarian approach – an approach not conditioned by the authority of the past. What is necessary is the exploration on behalf of the human race.

When the Subject Cognizance is in direct communion with nature, or in direct perception, the nature of the spiritual, energy flows to the object. The subject of cognizance is in non-dual and non-conceptual realm, there is no duality. The spiritual energy is beyond the energy of thought-ego, beyond the sentiments, emotions and other psychological imbalances, beyond the emptiness or silence of the mind.

Objective science becomes fulfilling when it imbibes the subjective division. Subjective science and spirituality are complementary. Both deal with the wholeness of life. This helps to evolve a complete civilisation for which the world is ripe and waiting.

In spirituality, we proceed from the awareness of the whole and with the awareness of the whole we proceed to the particular. Spirituality asks the question – "What are your instruments of perception?, act of perception, and the quality of consciousness behind the perception.[29]

The potential energy concealed in the human being – the energy of Intelligence – is independent of rationality. The creativity – *intuition, invention and discovery* – is a parallel brain; it rejects duality, rationality and psychological time.

Mankind needs freedom; trans-psychological need of love; an inner non-verbal, non-rational urge for love and peace. Man has to achieve the symbiosis of science and spirituality.

9. What is Consciousness?

Consciousness performs a dual role in the universe. In the world of the quantum, it is both the awareness and the creation of experience. It is the *being* and *knowing* of *experience.* Wilder Penfield, a renowned Canadian surgeon, wrote, "Our minds and our brains are not at all the same, it is the mind which experiences and it is the brain which records the experience".[30] He added, "To suppose that consciousness or the mind has location is a failure to understand neurophysiology".[31] Fred Alan Wolf wrote, "We are beginning a new age of consciousness, of the age of conscious atom. By looking within ourselves, we may be able to solve the problems facing us on the final frontiers - the frontier of human spirit".[31]

Larry Dossey, an eminent medical practitioner, wrote, "Nonlocal mind was infinite in space-time – Consciousness is both trans-spatial and trans-temporal; it is not in space and time. Consciousness is nonlocal and infinite, therefore is mortal and eternal". 'Nonlocal' is another word for 'infinite' – for if something is nonlocal or infinite in space; and if nonlocal or infinte in time, it is eternal and immortal.[32]

The time-space matrices which we experience is hierarchical quantum energy potential of consciousness in that frame of reference of mind spectrum. George Wald wrote, "The very crux of the twentieth-century physics in the recognition that the observer cannot be excluded from his observation, he is an intrinsic participant in them".[33] D.S. Kothari wrote, "What is observed is different from what it was before the act of observation".[34] Eugene Wigner wrote, "The measurement is not completed until its result enter our consciousness .... [this] last step has been is, at the present state of knowledge, shrouded in mystery and no explanation has been given for it so far in terms of quantum mechanics, or in terms of any other theory".[35] Hideki Yukawa wrote, "it is obvious that unless something rises into the consciousness it cannot become an object of rational thought, of rational consideration".[36]

To sum up the recalled subject of cognizance undergoes coherent superposition on the object, resulting into distortion of both the subject and object; we are aware of the distorted subject as modified memory and distorted object as distorted perception.

Scientific Research and Temper: When there is "Objective observing and understanding", which is without identification, without evaluation, without comparison, without measurement, without verbalization, the hierarchical level of potential of awareness/consciousness enhances; and the psychological mode of mind makes a quantum jump to scientific research and temper.

No-Mind State: If this observation and understanding is continued for a long chronological time, the mind goes to abeyance and we are left with a state of "no-mind". This is also called the state of silence. The silence is the emptiness of your consciousness when all this paraphernalia is brushed aside. The mind-brain-body system is totally free from the "I", the "me", the "ego".

Energy of Intelligence: The state of "no-mind" blossoms to the regeneration of Intelligence, or Wisdom, or "benediction of the timeless". "Meditation" or symbiosis of "Spirituality and Science" is a way of living where relationships are movements of relaxation and not tensions and pressures. Human beings are born of enjoying unconditional inner freedom. It is only free individuals who can cooperate with one another.

10. Concluding Comment

The immutable laws of Cosmic Intelligence or Nature are: (a) it is full of love and compassion, of non-violence and non-conflict, of harmony and order, and shares and sustains consciousness of all existences; (b) everything in the cosmos is a construct of the conditioned human mind, including psychological imbalances, acquisition, fear and so on; and the world extended in space and time is but our representation; (c) all floods, catastrophes, calamities, earthquakes, are not due to nature's wrath against mankind, but are constructs of the conditioned mind-brain-body system or inheritance of man over uncountable aeons; and (d) one is deeply struck by his concern about the large gap between the superhuman task [to embrace creativity, insight] and the frailties of human endeavours.

Albert Einstein wrote, "Our task must be to free ourselves from this prison by widening our circle of compassion to embrace all living creatures and the whole nature in its beauty". He added that we must live in a world of "objective observing and understanding" which leads us to attain "the peace and serenity". It is emphasized that "observing and understanding are embedded in a non-dual and non-conceptual frame of Mind Continuum.

We have to restructure science so that it takes in its realm both the objective and the subjective dimensions of life. The restructured science has the dimension:

[Science – Education – Ethics – Health]

The ethics imply to take in its ambient the sensual and trans-sensual dimension of life. A living in the *present,* the *now* heals and living in the *past* and *duality* suppresses man's perspective sensitivity and enhances disease. We must be aware of the mutation of the whole. It is the higher level of awareness that heals.

We have to jettison human intellect and reason, its information [knowledge] and experience and imbibe intuition. The *intuition, invention* and *discovery* belong to the expressions of nature.

There is no conflict between meditation and symbiosis of science and spirituality. Both have the identical aim of discovering *truth* and helping man to grow physically, mentally and spirituality, and achieve fulfillment. These are two complementary disciplines in the life of man will produce fully integrated human beings, and thus help to evolve. The meditation is a religious pursuit, while symbiosis of science and spirituality belong to the "scientific mode and temper". At the level of expression of "healing energy" or "energy of Intelligence" both the approaches became one and merge together.

Understanding comes when the cerebral activity goes to abeyance. The human being seems to be born with the faculty of self-awareness and a hidden energy of Intelligence. The understanding which is not knowing has a perceptive activity.

Man urgently requires a psychic mutation in his life. This leads to a genuine urge to explore the meaning of life, faith in life, the nature of truth and nature of Cosmic Intelligence. One may get into an orbit of human consciousness, a spark of love and compassion.[37] This is an enquiry on behalf of the whole human race.

References

1. Spencer B. Brown, *Laws of Form,* New York: Julian Press, 1972, pp. 104-105.
2. J.W.N. Sullivan, *The Limitations of Science,* New York: Mentor Books, 1949, p. 10.
3. E.N.da C. Andrade, *An Approach to Modern Physics,* New York: Doubleday Anchor Books, 1957, p. 255.
4. Albert Einstein, *I Believe:* Nineteen Personal Philosophies, Unwin Books, London, 1962, p. 26.

5. Schopenhauer, in Albert Einstein, *ibid.,* p. 26.
6. Discussion Jawaharlal Nehru and J. Krishnamurti, in Pupul Jayaker, in *Krishnamurti: A Biography,* San Francisco: Harper & Row, Publishers, 1986, p. 397.
7. J. Krishnamurti, in Pupul Jayaker, *ibid.,* p. 392.
8. *Ibid.,* p. 379.
9. *Ibid.,* p. 380.
10. *Ibid.,* pp. 386, 387.
11. *Ibid.,* p. 398, 399.
12. *Ibid.,* p. 433, 436.
13. *Ibid.,* p. 435, 371.
14. Albert Einstein, *Albert Einstein: 14 March 1879-18 April 1955,* Nature, Vol. 278, 15 March 1979, p. 215.
15. Galileo Galilei, in Hediki Yukawa, *Creativity and Intuition,* Tokyo: Kodansha International Ltd., 1973, pp. 110-111.
16. Voltaire, *Dictionnaire Philosophique*, Paris: Garnier, 1954 in Ilya Prigogine and Isabelle Stengers, London: Bantam Books, 1984, p. 257.
17. Herman Weyl, in Ilya Prigogine, *et al,* p. 311.
18. du Bois Reymond, in Ilya Prigogine, *ibid.,* p. 77.
19. Ilya Prigogine and Isabelle Stengers, *ibid.,* p. 6.
20. Hideki Yukawa, *op. cit.,* p. 204.
21. Erwin Schrodinger, *What is Life? & Mind and Matter,* Cambridge University Press, 1967, p. 134.
22. Hideki Yukawa, *op. cit.,* pp. 206, 101.
23. Max Born, in Hideki Yukawa, *op. cit.,* pp. 189-190.
24. J. Krishnamurti & Prof. David Bohm, *The ending of Time,* San Francisco, Harper and Row, 1985, pp. 105, 107, 110.
25. On J. Krishnamurti, *Within the Mind,* KFI, 1982, Preface, p. 6.
26. Sampooran Singh, *Scientific Research and Temper,* Snow & Aualache Study Establishment, R&D Development Centre, Chandigarh, January 29, 2007.
27. *Ibid., Creativity and Innovation, ibid.*
28. *ibid., A Vision on Tsunamis, Earthquakes, Snow & Avalanches and Global Warming,* CASE – 2007, 2 March, 2007.

29. Vimala Thakar, *Science and Spirituality,* Ahmedabad: Vimal Prakashan Trust, 1988, pp. 1-39.
30. Wilder Penfield, in Deepak Chopra, *How to Know God,* London: Rider, 2000, pp. 215, 217.
31. Fred Alan Wolf, in *Taking the Quantum Leap,* San Francisco: Harper and Row, 1981, pp. 244, 248.
32. Larry Dossey, in *Vision and Values,* Lalit K. Kothari and Ramesh K. Arrora, New Delhi: Paragon International Publishers, 2006, pp. 49-56.
33. George Wald, in *Synthesis of Science and Religion, op. cit.,* 19.
34. D.S. Kothari, *Some Thoughts on Science and Religion,* the University of Delhi, 31st October, & 3rd November, 1977, p. 25.
35. Eugene Wigner, in D.S. Kothari, *ibid.,* p. 26.
36. Hideki Yukawa, *op. cit.,* p. 203.
37. Vimala Thakar, *Through Silence to Meditation, op. cit.,* 1998, p. 49.

Questions and Answers

Q. 1: What is the Cause of the Conditioned Psyche, or the Psychological Mutation, or the Transformation? How do we Resolve this Paradox?

Ans.: Without being aware of it and without being systematic about it, the human mind made an error of constructing in uncountable aeons the space-time matrices – the world extended in space and time is but our representation. P. Krishna stated, "Has humanity taken a wrong turn which has created this chaos in the world?" and "What needs to be done within the consciousness of man to preserve our very existence and bring about a new culture?"

Man has tried to purify the mind for more than a million years, but has utterly failed. We are busy with the psychological mutation. We are missing the communion with the real because we are ever busy with the unreal, the non-factual.

The answer to this paradox is that man has to undertake an enquiry, an observing the meaning of life, the facts of psychological life, the trans-psychological truth behind these facts, and we enquire the absolute ground of our being, the ground of our existence. The only concern of an enquirer is to understand the truth, in order to be able to discriminate the true from the false. It is emphasized that an enquirer [observer] is synergistic to the conditioned psyche, or in other words, it is at a higher quantum energy potential of awareness than the conditioned mind and the subject of cognizance makes a quantum jump to a higher level of awareness which is shared with the object. You live in the light of the truth that you understand. This is the blessing of living or the benediction of nature.

Q. 2: Please Discuss Knowledge, Learning and Awareness?

Ans.: The conditioned mind gathers experience and information [knowledge]. Knowledge is a part of consciousness, as a part of your inheritance; and awareness [consciousness] is the principle of life. Awareness is not the method of enquiry. Enquiry is possible in a non-dual and non-conceptual frame of reference of Mind Continuum. It is the by-product of learning. Enquiry is not an acquisitive activity. The knowledge gets transformed to memory and is recorded in the brain; it becomes a part of the thought-structure. Acquisition, possession, ownership – the whole paraphernalia belongs to acquisitive activity. Learning requires the commitment of your whole being, it requires the openness to life. So learning gets implemented in the movement of life effortlessly. Learning leads to living. When a person thus enquires, the person learns and awareness is the perfume of that learning.

21

Nature's Benevolence of the Hidden Potentials of Consciousness

1. Introduction

We conduct an enquiry of the meaning of life, the facts of psychological life, the transpsychological truth behind those facts and we enquire into the absolute ground of our being. The only concern of an enquirer is to understand the *truth,* in order to be able to discriminate the truth from the false. You live in the light of the truth that you understand.

Mutation is a sudden, abrupt, qualitative change in the totality of our consciousness. This article does not discuss the biological mutation as it happens in a part of the biological structure. We discuss the psychic mutation, which happens in the whole of the mutant. Psyche implies consciousness and the consciousness through which we live, with which we perceive, with which we respond in relationship, is daily life, in conditioned.

The first thing should be to learn to watch, to observe, to look at the movement of mind incessantly, defencelessly, non-aggressively. In the act of observation, we live in total freedom.

Maurice H. Wilkins, Nobel Prize in Medicine, 1962, wrote, "An open-minded-enquiry of the scientist is not something peculiar to science itself but a characteristic of the good way of living for human beings in general. It is equivalent to the religious concept of love, where you are always giving attention to new developments and new possibilities. To say that the essence of science is that you are always enquiring and open-minded is to say that you are in fact living a virtuous life". An openness or receptivity implies that inwardly we are not pre-occupied with any other than the action of perception, the action of audition, listening.

Man is an amphibian who lives alternately in two worlds, firstly, man-made social cosmos or relative frame of reference of mind construct, and secondly, man lives in a creative cosmos, or a construct of nature, or absolute rest frame of Mind in the Mind Continuum. As an analogy, we may compare the man-made cosmos to a drop of water of an ocean, and creative cosmos to an unfathomed unbounded, limitless ocean. This suggests that we have to learn to jettison the symbolic-dualistic frame of reference and learn to live in the nature's creative cosmos according to the immutable laws of nature.

This article discusses nature's benevolence to regenerate the hidden potential of Consciousness. The essence is Albert Einstein's prophetic words, "Objective observing and understanding", which leads to "peace and serenity". He added, "Our task must be to free ourselves from the prison by widening our circle of compassion to embrace all living creatures and the whole nature in its beauty".

2. Albert Einstein and Peace

Albert Einstein, *"Out of His Later Years"* perceived that the conditionings had increased at an accelerative pace. He wrote, "Nothing of all that will remain but a few pitiful pages of the history books, briefly picturing to the youth of future generations the follies of its ancestors".

In 1932, after a World Disarmament Conference in Geneva, Einstein, in an unusual bout of anger, exclaimed, "they (politicians and statesmen) have cheated us, they have fooled us. Hundreds of millions of people in Europe and in America, billions of people have been cheated, traded and ticked out of their lives, health and well being". These remarks are equally valid today. In principle man is deceiving himself.

Albert Einstein stated, "It is in fact, nothing short of a miracle that the modern methods of instruction have not yet entirely strangled the holy curiosity of enquiry, for this dedicate plant, aside from stimulation, stand mainly in need of freedom, without this it goes to wrack and ruin without fail".

Einstein wrote, "Humanity needed to reach a higher level of awareness to solve the problems of nuclear war". Martin Luthar King rightly said, "Science has completely finished our spiritual power, we have guided missiles but misguided men".

3. A Critique on "Objective Observing and Understanding"

When a system with higher quantum energy potential observes another system with a lower quantum energy potential, the lower one is knocked off or goes to abeyance. This bestows understanding and learning. The intuition (insight), spontaneity and right action is born in the Mind Continuum which blossoms to regeneration of Supreme Intelligence and bring in mutation of the conditioned human psyche.

We will have to initiate educating our physical organism and psychological structure, and work for a paradigm shift in the whole system of education. We will have to work for restructuring of science, restructuring of education, awareness of human perennial values, and live that understanding every moment.

The real science (restructured science-education-health) and real religion (restructured Religion-Faith-Values) are in search of truth and both approaches are non-authoritarian, non-dualistic and non-causation. We urgently need a symbiosis of real science and real religion, which are the dimensions of the wholeness of life, and as such are synonymous. Every existence/ event/process/ happening is permeated and surrounded by the Cosmic Consciousness, the Life Field, the Unmanifest.

By offering one's life for that psychic mutation to take place, one might create a living cell of a new life style, new way of living. One might be setting into the orbit of human consciousness a spark of love and compassion. One embarks upon that enquiry, not for oneself, but on behalf of the whole human race. This is fundamental to the survival and excellence of man. This is the consummation of human life and living.

To sum up, observation leads from knowledge to understanding; from understanding to anticipation of a new energy which is called energy of Awareness, that has no centre, no past, no conditionings; this is also called energy of Intelligence, energy of pure Consciousness, or Cosmic Consciousness. The dimension of silence – it is non-personalized energy, non-individualized energy that heals the distorted psyche and ends all imbalances.

4. Understanding of Life Directly

Nature out of its benediction has provided man a hidden potential – known as non-dual and non-conceptual frame of reference of Mind Continuum; or in other words, an intuitive mode of mind. Many eminent scientists – including Albert Einstein, Werner Heisenberg, Erwin Schrodinger, Arthur Eddington, and so on – have stumbled upon another made of perception (knowing), which is understanding of life directly, instead of in the abstract, linear terms of representational thinking. This implies understanding the wholeness of life. **Understanding is a movement in the non-duality, a movement of the whole being.**

Understanding comes only when the mind is very quiet, tranquil, silenced. It leads to perception of the whole. Perception has nothing to do with knowledge (information). **The perceptions are born of non-duality and responses are born of spontaneity.** It awakens, the awareness, the intelligence, which arises out of observation, self-knowing [understanding] and compassion. It implies that ultimate perception happens in a trans-sensual dimension, which is timeless [beyond brain]. As all knowledge is structured in Consciousness, so ultimate perception is a happening when Universal Consciousness bathes the brain cells. Perception bestows understanding the wholeness of life which flowers to intuition and right action; it is a happening in the non-dual and non-conceptual frame of reference of Mind Continuum.

5. Meditation

Meditation is not a state; it is a movement, an action is a movement. If we bring in observer between the learning and the action; then the learning stops. To learn there must be complete freedom. The word whole not only means healthy but holy.

It is only the action of will and assertiveness that needs to be changed at all, because the only mischief in relationship is conflict, between individuals or within individuals, and conflict is will and assertiveness.

For a qualitative change in consciousness, the austerity of learning, discovering, understanding and living your understanding is necessary. We have to jettison the *non-fact,* the *past,* the *false,* the *what should be?*; and imbibe the *fact,* the *present,* the *true,* the *what is?* The necessity today is to break away from the traditional, conventional way of psychic living and be left with life – to be left with the emptiness of consciousness, which is the choiceless encounter with life and be willing to let the life operate upon us.

The symbolic-dualistic frame of reference has the lowest quantum energy potential of consciousness [conditioned energy]; in observation and understanding there is energy of Attentivity; and at the highest quantum energy potential of Consciousness [higher frequency means higher energy], and this is called energy of Intelligence.

6. Quantum Jump of the Conditioned Psyche to the Supreme Intelligence?

Psyche implies consciousness and the consciousness through which we live, with which we perceive, with which we respond in relationship, in daily life, is conditioned. The thought-structure coupled with the "I-consciousness" [the "me", the "ego"] have constructed the conditionings.

The accelerative progress of science and technology created objective pressures and compulsions that followed the growth of global politics and global economics. Man has lost touch with the facts of his life and he is still left on the mental and intellectual plane with the theories and ideologies, conclusions and objective values. He is not at peace with himself, and torn within himself. He lives in constant tension, violence, conflict and contradictions. Collectively also man is not at peace.

We have to turn the searchlight of attention [awareness, consciousness] on our own conditioned psyche. We have to acquire a higher potential of consciousness to understand oneself without the intervention of the conditioned mind; and this is called, silence.

"Silence is gathering of energy upto itself, which implies that silence is a way of living which is at a higher hierarchical level of quantum energy potential of consciousness". Silence is a dimension of living at the objective and the subjective in one sweep. In the sense of freedom – the new dimension of consciousness grows, where there is direct communion, direct understanding, pure perception.

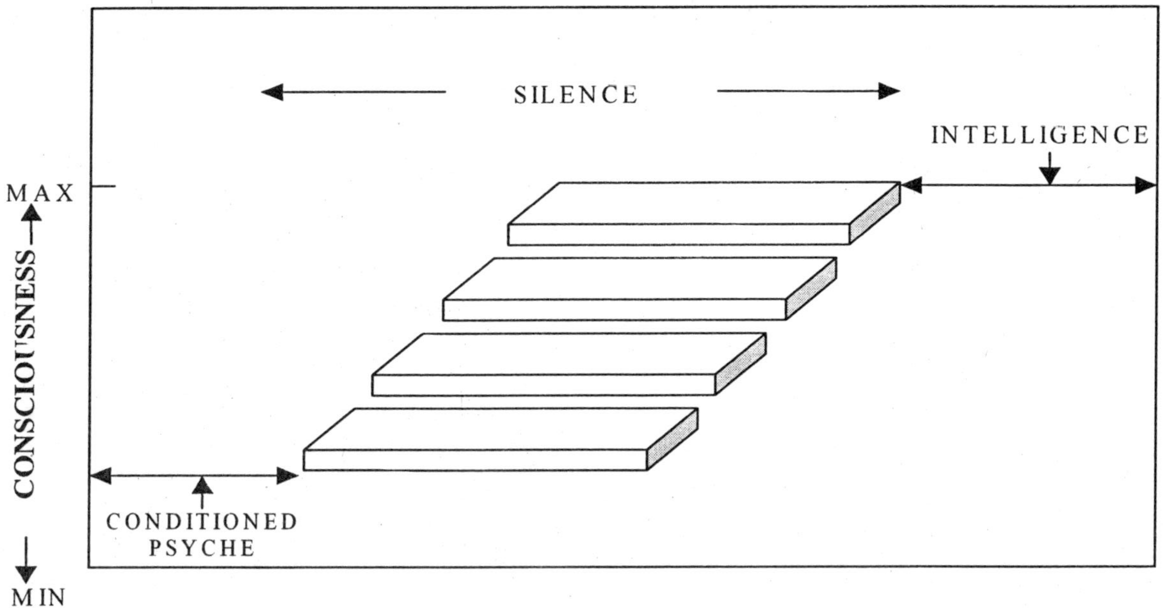

Fig. 3: The whole process of conditioned psyche is conceptualization, ideation and identification. The way to transcend the conditioned psyche is the way of silence, the way of meditation; it is a bridge between the conditioned psyche and the supreme potential of Intelligence. Intelligence is to grow to a non-subjective, non-rational dimension of consciousness.

The way of silence or the meditative way of living is a bridge between the conditioned psyche and the Energy of Intelligence. Fig. 3 shows that the way of silence is synergistic to the conditioned psyche. If this enquiry is continued for a long chronological time the Energy of Intelligence is awakened.

The way to transcend the conditioned psyche is the way of silence, the way of meditation. The way of silence is allowing the mind to cease to function – a state of no-mindedness. It is state of cessation of mental activity. It is total relaxation of the whole mind. Silence is looking at things as they are. It is awareness of the *fact*, the what is. This is expressed as under:

[Silence] →[Conditioned psyche] →Intelligence

Silence becomes a pathless path for a non-cerebral enquiry. Silence is at a higher awareness than thought, emotions and re-actions.

The truth about anything can be known only by awareness. If the enquiry is sustained, the energy of Intelligence gets activated. This is the highest quantum energy potential of Consciousness. It resolves all psychological problems, including violence, strife, fear. Life, the supreme master, is creative energy and therefore life, being inexhaustible creativity and infinity, immeasurableness, learning has no end.

Mahatma Gandhi wrote, "Truth is my God, truth is my defence, truth is my strategy". He had put something into the orbit of global human consciousness and even today what he had set into motion, is still alive. The human survival and excellence depends on moving with life, sharing its creativity.

7. A Critique on Education

(a) David Bohm, an eminent physicist, wrote that we need a new "instrument of thought", which would radically alter the consciousness of the observer, reorientating it towards a perception of the "unbroken wholeness" of which everything is a form. He added, "We must turn physics around. Instead of starting from parts and showing how they work together, we start with the whole."

It appears that physics must start with the whole [subjective dimension of life], and then explain the objective parts with the light of the Wholeness. The inner dimension is the whole, while the outer dimensions are the parts. It is the wholeness that heals, where the healing implies ending of all psychological imbalances, and there is perfect freedom and peace.

(b) **Consciousness** : Erwin Schrodinger wrote, "Consciousness is never experienced in the plural, only in singular... Consciousness is a singular of which the plural is unknown." Universal consciousness is the Wholeness, or Oneness. And everything that exists is an expression of consciousness at lower quantum energy potentials.

8. Concluding Comment

G. Spencer Brown, in a most spectacular passage wrote, "In order to see, evidently it must cut itself up into at least one state that sees and at least one other state which is seen. In this mutilated condition, whatever it sees is only partially itself". The object recalls the subjective [recalled memory] and there is coherent superimpositions of different webs with the result that both the object and the subject get deformed. The world in seen not as it is, but in its various relationships to the "me" of the memory. This gives us experience and information [knowledge] where both are the *non-fact,* the *past,* the *false,* the *what should be?* This leads mankind to violence, aggression and self-annihilation.

The human-made social cosmos or construct of the symbolic dualistic frame of reference of mind spectrum represents the immanent quantum energy potential (successive consciousness) and is divorced from the ultimate quantum energy potential, Simultaneous Consciousness. **The construct of the mind is the only "reality" where the transcendent potential has gone to abeyance and its only linkage is the immanent potential. It represents the relative field of consciousness.** All activities in the relative frame of references are antagonistic to the laws of nature.

Nature out of its benediction has provided man a hidden potential – known as non-dual and non-conceptual frame of reference of Mind Continuum; or in other words, an intuitive mode of mind. All natural processes – seeing is direct, hearing is direct, touching is direct, but thinking is indirect. All great discoveries have arisen from the realm of the timelessness, so scientific enquiry is not a cerebral activity and gathering information, but a quest of the Timelessness, the Divinity, the Life field. A selfless, unfaltering devotion to truth and use of knowledge of man's uplift – higher hierarchical expression of consciousness – are the only vocation of a scientist.

We have to jettison reason and intellect; the experience and information [knowledge]; and imbibe intuition. The *intuition, invention* and *discovery* belong to the direct expression of nature. Nature, in its benediction, is full of harmony and order, non-violence and non-conflict, love and compassion; perfect sustainable freedom and peace. A man with right "observing and understanding" is free from political activity, religious conformity and tradition. It implies that he is free from the weight of time, from the burden of the past, free from all the action of will. This only is the social, psychological and even the political revolution.

Glossary

Mind	Mind [with a lower case of 'm'] is a conditioned energy. The mind constructs symbolic–dualistic frames of reference of mind spectrum; parts of the map, not of the territory.
Mind	Mind [with upper case 'M'] is non-dualistic and non-conceptual Mind Continuum.
Mind-Brain-Body system	It is the mind which experiences and it is the brain that records these experiences.
Perception	Seeing is direct, hearing is direct, but thinking is indirect. The Mind Continuum has the capacity for pure perception.
Violence Non-Violence	It is a construct by the symbolic-dualistic frames of reference of mind spectrum. The non-violence is a percept in the non-dual and non-conceptual frame of reference of Mind Continuum.
Creativeness	The intuition, invention, discovery and intelligence belong to creativity.
Innovation	Innovation is defined as a movement of "quantum energy potential of consciousness" from an unfinished process/product to a commercially useful product - as the final product is at a higher order of consciousness than the unfinished product.

Objective Observing and Understanding	The "objective observing and understanding" leads to quantum jump of consciousness from dualistic frame of mind spectrum to non-dual and non-conceptual frame of Mind Continuum.
Healing Mutation	When the conditioned mind is bathed by Consciousness, all conditionings end and this is called healing or mutation of the psyche. The healing enhances the R&D potential of production of commercial goods.
Jettison	It is to throw (goods) over-board in order to lighten a ship.
Quantum Jump	It is the tiny but explosive jump that a particle of matter undergoes in moving from one place to another.
Ego, 'me'	The ego has no factual reality at all. It has a conceptual reality. It has an ideational reality and it has a relevance to our lives. The thought-ego system cannot be separated.

Index

D

E

J

K

L

M

N

Q

R

S